OFFICE TECHNOLOGY

ED SOBEY

Library of Congress Cataloging-in-Publication Data
Sobey, Edwin J. C.
A field guide to office technology / Ed Sobey. — 1st ed.
p. cm.
Includes bibliographical references.
ISBN 978-1-55652-696-1
1. Office equipment and supplies. I. Title.

HF5521.S575 2007
651'.2—dc22

2007006763

Cover and interior design: Joan Sommers

First edition
Published by Chicago Review Press, Incorporated
814 North Franklin Street
Chicago, Illinois 60610
ISBN-13: 978-1-55652-696-1
ISBN-10: 1-55652-696-2
Printed in the United States of America
5 4 3 2 1

To our long-term Florida friends—Tim and Mary, Julie and John, Ray, and other diving and museum friends. We think of you often, especially in the drizzle months.

CONTENTS

ACKNOWLEDGMENTS

Because I hang out in museums, many of the photos I took are from the offices in museums. Thanks to Putter and staff at KidsQuest for allowing me to take photos there. And, thanks to the staff of the Elmhurst Historical Museum in Illinois for giving permission to take photos.

Steve McCracken, my personal telephone trainer and guru, again helped me fathom the mysteries (he assures me they aren't mysteries) of the telephone system. Thanks, Steve.

Frank Handler, a friend and fellow Fellow of the Explorers Club gave me a guided tour of his office so I could take a bunch of photos. Chris McArthur allowed me to use his photo of a laser pointer he had taken for OnPoint Lasers, Inc.

Mark Hovde let me photograph his high-tech toys (GPS), but didn't let me play with them. Rod Brown led me through his offices on a photo expedition that snagged some important shots. Bob Spiger helped me with information about building codes and fire alarms.

Andrew Oh Boon Keng at the Singapore Science Centre provided the image of the Segway. Michael Weizer, director of marketing at ACCELL, provided the photo of the USB light. Tony Tang and Jonathan Clark provided photos of docking stations. HP allowed us to use the image of the digital pen—very cool technology. The smart window photo is used with permission of Lawrence Berkeley National Laboratory. Paul Singleton allowed the use of his photo of a video phone. Bob Wismer, ace runner and run organizer, took the photos of the PBX and time clock. The flushless urinal photo is courtesy of Waterless Co. LLC. The photo of the time clock was provided by Acroprint Time Recorder, Raleigh, North Carolina. John Dickson came through again with some great photos of devices. Thanks, John. On Point Lasers gave permission to use a photo of its laser pointer and wrote a very nice note to me as well. Thank you kind people.

Tuscaloosa County School System allowed me to use its photo of its PBX. SAGE Electrochromics, Inc. provided several great photos of its electronically tintable windows and explained how they work. My brother-in-law, John, once again assisted—this time by taking photos of the parking lot sensors.

While attending Central Washington University, Justin Mumm took several of the photos I have used. Thanks, Justin, and good luck in your career in photography.

Jerome Pohlen at Chicago Review Press has been great to work with; his keen suggestions have made this work and the previous two *Field Guides* much better than I could have hoped. I also thank Michelle Schoob for her great contributions in catching my mistakes.

INTRODUCTION

Finishing up my graduate studies in the 1970s, I wrote my dissertation in longhand and took it to a typist. She had the thankless job of converting my page scrawling into what I meant to say. When I got the manuscript back from her, I read it line by line, making marks both where I wanted to make changes and where she had misread my writing. Then, she started over at page one and retyped the entire manuscript—153 pages' worth. I checked the new manuscript line by line for any errors or changes. After a few more minor corrections, I took the pages to a print shop where they were printed and bound.

That was the way documents were created back in the Predigital Age. At the time, we thought we were marvels of technology. Each company, each department on campus, each organization had one (or more) large rooms filled with secretaries typing away. Gone were manual typewriters, replaced with electrics.

Copy machines came into their own in the late 1960s and early '70s. As a naval officer at sea in 1971 I used a thermal copier. You placed the page to be copied against a wax paper-like film and ran it through the machine. Then you took the output of that process and ran it through the machine a second time with a piece of special copy paper. You could copy a page a minute. Who needed carbon paper?

The office has changed dramatically in the last few years. The once common secretarial and typing professions are exceptionally rare.

Machines have replaced people doing data entry, filing, finding, calculating, and more. More gizmos are invented every year, and some wiggle onto our already crowded desktops.

Today I am typing this book into my desktop computer. No handwritten drafts and no secretaries trying to read my handwriting. I'll run the spell checker to ensure that I haven't shown my publisher how poorly I spell; and after reading and making changes to the manuscript, I will (the computer will do all the work) copy it onto a CD. My heavy lifting is dropping the CD (and hard copy) into the mail to Chicago Review Press.

The bottom line, the thing everyone wants to get to, is that we are more productive today. We do more work with less effort, which sounds great to me. (Still, we aren't working less. The short work week predicted a few years ago has not materialized, and instead of having too many workers and not enough work, at least some industries have labor shortages. Go figure.) Where is it all leading?

A few years ago the experts were predicting the "paperless" office. No one would use paper, they opined, as everything would be stored and retrieved digitally. This is yet another example of the futility of making predictions. One prediction I feel comfortable making, however, is that the office will continue to be a place of changing technology.

In a place that changes so rapidly you need a guide to "What's that?" Even if you know what it is, do you know how it works or where it came from? If you work with a device every day and are the least bit curious, *A Field Guide to Office Technology* belongs on your desktop. You can store it where you used to keep those 5 ¼-inch floppy disks.

As you wander around the office, take this book with you. If you spy something strange and tech-like on the wall, you can now find out what it is. Thumb through the On the Walls chapter and compare the photos to what you're looking at. The entry will tell you what the device does, its "behavior"; where it is typically found, its "habitat"; how it works; and other interesting facts. With a little office exploration and reading you could be out-teching the tech guys. How cool is that?

So keep this guide handy. Reach for it whenever you hear the words "What's that?" And have fun exploring the office.

ENTERING THE OFFICE

A FEW YEARS AGO I was dropping off a proposal at Paul Allen's office. I knew it was unlikely that I would see him, but I was looking forward to seeing what the (outer) office of one of the world's richest men looked like.

I walked down the corridor and was 20 feet from his office door when a voice greeted me. The voice was from somewhere above or behind me; I couldn't tell where.

"Can I help you?" asked the faceless voice.

Not sure who was talking or who the voice was talking to, I continued toward the door. But then the voice repeated itself.

I looked around again and not seeing anyone, mumbled, "I have a package to drop off."

"Deliver it to room 200, down the hall and one floor down," directed the voice.

Not only was I prevented from seeing the great man, I couldn't even see his office or the person behind the mysterious voice. I couldn't get in the front door.

Just getting into an office today involves technology that didn't exist a few years ago. Of course some technology, such as keys and mechanical locks, has been around for centuries. But parking lot magnetic card readers and remote control door openers are fairly new. They provide a high level of security and certainly a dose of awe.

If you're able to get into the office, another layer of technology awaits you. An emergency enunciator mounted in a wall is ready to tell incoming fire and rescue workers where the trouble is. A time clock may be ready to tell your boss where you are . . . or where you aren't. Security control boxes and keypads allow you to enter your office and punch in a code to keep the alarms from going off. Automatic door closers and exit lights help keep us safe.

So don't just walk into an office; play the sleuth. Take another look at the stuff in plain sight that you've seen hundreds of times before not knowing what it is or what it does. Then search this chapter to find the answers.

Garage Door Opener Touch Pad

BEHAVIOR

Allows you to open a parking gate or garage door without a handheld remote control unit kept in your car. System managers can change the code to keep out people who no longer warrant access.

HABITAT

Touch pads are mounted on metal poles or stands immediately in front of the gate, so you and only you can drive in when the gate opens. The same type of device is used outside office doors to keep people out of restricted areas.

HOW IT WORKS

When you punch in your code, it is recognized by a logic circuit in the garage door electronics and opens the door. The motor that opens the door is started by the adjacent circuit. Typically the door or gate closes after a fixed time period.

For vertical doors, the motor doesn't lift the entire weight of the door, which can be enormous. Springs counterbalance the door's weight, so the motor does less work. In sliding gates, the weight is supported by wheels that run on rails or on the ground.

INTERESTING FACTS

Some systems allow managers to record entry into the garage or lot, so they can tell who is on the premises at any time.

Parking Lot Sensor

BEHAVIOR
Opens the gate or door so you can leave the garage or lot.

HABITAT
The sensor is installed in the ground in front of the exit lift gate or door, or installed two car lengths in front of the gate.

HOW IT WORKS
Some garages require you to use the same remote control to leave that you used to enter. However, most employ some type of sensor to detect the presence of a car or motorcycle (but probably not a bicycle—you're out of luck unless the garage provides a manual push button).

One approach to opening the gate is to bury a coil of wire in the ground and run an electric current through the coil. The coil generates an electromagnetic field. When a car drives on top of the coil, it changes the magnetic field and the change is detected by a sensor. The exit door or gate system continuously checks the sensor to see if the field has been changed. When the sensor "sees" the car (a changed magnetic field), the system starts the motors that raise or move the gate.

Some systems use an infrared beam to detect cars. You can see the beam transmitter off to the side of the gate or door. It has an eye-like device that allows infrared radiation to pass. It may have a second eye-like glass cover where the beam, reflected from a silvered device on the opposite side of the driveway, bounces back to it. This two-eyed system is less expensive because it doesn't require installation of a sensor on the other side of the driveway.

In the alternative system, the detector is located on the other side of the driveway. Cars waiting to leave block the beam to trigger the gate to open.

UNIQUE CHARACTERISTICS
In-ground cables for the sensor are usually installed after the lot has been paved. So look for a black square or circle in the ground where the sensing cable was installed and covered with hot tar.

Security System Keypad

BEHAVIOR

Allows workers to enter the office without setting off the alarms. Workers enter a code to arm (when leaving) and disarm (when entering) the office.

HABITAT

The keypad is located inside the entry door used by staff. It is inside to protect it from tampering. Being inside requires that it have a programmed delay—about 30 to 45 seconds—so workers can enter and disarm the system by punching in a code before it triggers an alarm.

HOW IT WORKS

The 12-button keypad contains switches that are activated by pushing. When you push a button, it closes a switch and signals the system microprocessor. When the programmed code is entered, the microprocessor activates or deactivates the alarm system.

Each individual can have his or her own code so the system can identify who set or disarmed the alarm.

INTERESTING FACTS

You might have noticed that keypads and calculators arrange the numbers in different sequences: calculators have the "1" in the lower left, while keypads have "1" in the upper left. Calculators replaced mechanical adding machines and followed their arrangement of numbers. Keypads were developed by Bell Labs to replace the rotary dial telephone. If Bell Labs had followed the calculator key placement, the alphabet would start in the lower part of the pad and end with "WXY" in the top row, so it opted instead to present the alphabet in order from top to bottom and left to right.

Front Door Lock

BEHAVIOR

Keeps out (some of) the people you don't want inside, but gives those you do want inside easy access.

HABITAT

Mounted on the side of the door opposite the hinges. Located at a convenient height to allow users to insert a key to open the door.

HOW IT WORKS

Most doors use a cylinder lock. Inserting the correct key allows you to rotate the cylinder, which is connected to an arm that withdraws the latch keeping the door secure. The latch is usually pushed closed with an internal spring after the key is removed.

The beauty of the lock is that each has its own code that protects it. The code is cut into a metal key. The vertical indentations in the key correspond to both the placement and heights of pins inside the lock. As the key enters, it pushes the spring-mounted pins up and out of the way. A key with a different code will not push the pins to the correct heights that allow the cylinder to turn.

From inside you can operate locks by turning a knob or using a key.

INTERESTING FACTS

The first lock system that used keys was created in Sparta around 400 B.C.

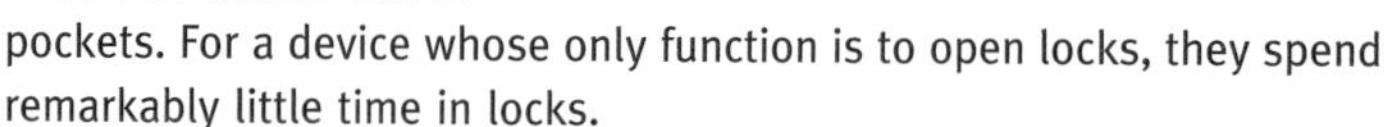

Key

BEHAVIOR

Gains access to locked doors and devices.

HABITAT

Keys hang around key chains or amidst lint in pockets. For a device whose only function is to open locks, they spend remarkably little time in locks.

HOW IT WORKS

Mechanical locks use keys that push internal pins out of the way. When all of the lock's pins are compressed by ridges on the key, the tumbler can turn, unlocking the lock.

UNIQUE CHARACTERISTICS

In the photo the key on the left is a tubular key. The center key is a traditional key, and the key on the right is a high-security key blank made by ASSA.

INTERESTING FACTS

There are many varieties of keys and locks. Here are some of the common types.

- **Tubular or axial pin tumbler locks and keys** are used occasionally on office doors, but more often on vending machines and bicycle locks.
- **Cylinder locks** rotate to open, after the key has been inserted. These are the most common door locks.
- ASSA and other **high-security keys** have unique shapes (the blanks are sold only to authorized dealers) that make it difficult to make unauthorized copies.

Magnetic Card Reader

BEHAVIOR

Allows entry to an office without a mechanical key. The magnetic card identifies who is entering so a record can be kept. In this world of downsizing companies, it is easy to reprogram the system to deny access to those who are left out in the cold.

HABITAT

Card readers are mounted on the walls outside the office doors or outside gates protecting garages and parking lots.

HOW IT WORKS

The plastic card has a film of magnetic material, like magnetic tape from a cassette tape recorder. The tape is plastic film embedded with tiny iron oxide particles that act like magnets. Electromagnetic encoders align the tiny magnets on the card so a read head can read them. If exposed to strong magnetic fields, the particles can be remagnetized into a new pattern, thus losing the code. In that case, you can't get in! Some systems also have a keypad for calling someone inside to open the gate or to punch in a code to gain entry.

In addition to magnetic card readers, security systems can use optically encoded cards (bar codes) or radio frequency identification (RFID) technology. RFID can be active (requiring a battery) or passive (using the energy of the incoming radio signal from the lock system) to power the onboard electronics circuits. Active RFID sends out a radio signal (up to 300 feet) identifying the owner so the receiver will open the gate. Passive systems are useable for a much shorter range.

Door Alarm

BEHAVIOR

When the alarm system is operating, it detects an opening door and sounds an entry alarm.

HABITAT

These devices are visually innocuous, and you may miss them. Look along the inside, upper edge of doors that open to outside the office. If you see a small device mounted to the door, look for a correspon-ding device mounted on the door frame. The device on the door frame will have a wire trailing out of it and running along the door frame.

HOW IT WORKS

The device mounted on the door is a magnet. Its corresponding piece on the door frame is a reed switch. The presence of the magnet pulls together two metal contacts inside the switch so current can flow through the system. This configuration is called "normally closed." When the door is open with the alarm system engaged, the system senses that the circuit is no longer complete and it signals an alarm. "Normally open" reed switches also can be used, in which case opening the door closes the circuit to signal the alarm. The wire running along the door frame connects the reed switch to the alarm system.

UNIQUE CHARACTERISTICS

You can sometimes find the same system on windows that slide open and shut.

INTERESTING FACTS

The reed switch is one of the many inventions of the old Bell Telephone Laboratories.

Security Panel or Alarm Control Box

BEHAVIOR

Controls the perimeter alarm system and may also control other alarms—motion, fire, and smoke detectors—throughout the office.

HABITAT

It is often located near the entry door or in a nearby closet, mounted on the wall. Typically, it is a gray-colored metal or white plastic box, but it can be painted to match the walls. In companies that have security guards, the alarms are located in or near the guard station.

HOW IT WORKS

A microprocessor inside the device receives the codes entered via a keypad by arriving or departing employees. On leaving at the end of the day, the last worker out enters a code to start monitoring alarms for entry. Once the correct code has been entered, the microprocessor waits a set period to allow the worker to leave and lock the door before activating the alarms.

The first employee entering at the start of the day will trigger perimeter and motion detector alarms, but the system will wait to sound the alarm, giving the employee time to disarm by entering the code.

System managers can set and change the code in the microprocessor. Every few months they should change the code, hopefully getting copies to all the employees who need it.

Some systems also monitor the fire and smoke detectors, or monitor sound levels inside the office. When any of these alarms is triggered, the system can send a message by telephone lines to a remote security company, or a police or fire department.

Entry Alert Device

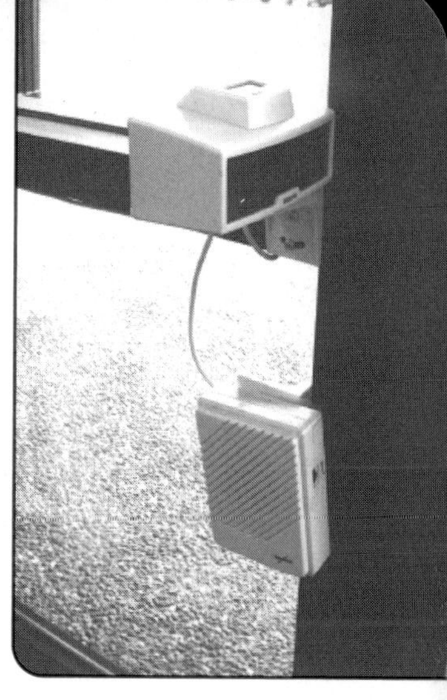

BEHAVIOR
People entering an office interrupt a beam of infrared light, and the device sounds a chime or tone to notify employees of their presence.

HABITAT
These are mounted a few inches above the floor on both sides of the entry door.

HOW IT WORKS
The system uses a microprocessor that sends a tone or chime to speakers whenever the beam of infrared light is interrupted. One device mounted on or near the doorway near the floor generates an infrared beam of light. A second device detects this beam. The second device is mounted on the opposite side of the entryway. Alternatively, both the light transmitter and receiver can be mounted one above the other, in one box. In this case a reflector (like a reflector mounted on a bicycle) is mounted on the opposite side of the doorway. The infrared beam crosses the doorway and reflects back to the receiver. When someone interrupts the beam, the circuit detects his or her presence and sends the tone to the speakers.

The microprocessor continuously looks to determine if the detector is "seeing" the beam. It queries the detector many times each second, so even a short interruption in the beam will be enough to trigger the tone.

UNIQUE CHARACTERISTICS
Look for the beam transmitter and receiver. Mounted in the same unit, they appear to be two eyes, one above the other. Look for the small reflector on the other side. If you missed seeing them when you walked in, your first clue to their presence will be the chime announcing your arrival.

Emergency Enunciator

BEHAVIOR

Announces emergencies and helps locate them for responding emergency crews. Imagine the fire department showing up at a huge building and not knowing where the fire is. This device shows where the alarms have been triggered.

HABITAT

Emergency enunciators are located in the entryway or security stations of office buildings.

HOW IT WORKS

When sensors and alarms in a building are activated, their location is indicated on the enunciator. Some systems also report mechanical problems with machinery in the building. In medical facilities, enunciators are used to alert on-duty staff to call buttons or wireless emergency buttons (panic buttons). Emergency responders can check the enunciator to find where the problem is.

UNIQUE CHARACTERISTICS

In any large building, look near the lobby, elevators, or front doors for a panel that lists locations throughout the building.

Door Closer

BEHAVIOR

Closes an office door without having it slam.

HABITAT

Door closers are located on the upper inside surface of most doors.

HOW IT WORKS

The closer is attached to both the door and the door frame. Opening the door extends the mechanical arm, which turns a mechanism inside the drum attached to the door frame. Releasing the door allows the closer to pull the door closed, resisted by a hydraulic piston.

Opening the door stretches a spring inside the mechanism. The spring, strong enough that you don't want to open the mechanism to look at it or attempt a repair, pulls the door shut. A piston filled with hydraulic fluid resists that motion, slowing the movement of the door. Struts or shocks and springs in your car work the same way.

You can adjust how fast the door closes with one or two small screws on the mechanism, but removing the screws may allow the hydraulic fluid to leak out.

INTERESTING FACTS

Lewis C. Norton invented the first door closer in 1877.

Electromagnetic Door Release

BEHAVIOR
Releases fireproof doors in the case of fires to prevent their spread.

HABITAT
In office buildings on doors separating different parts of the building. The releases are mounted on the wall near the top of the outside edge of the door. A corresponding cap attaches to the inside of the door.

HOW IT WORKS
When a fire is detected by a sensor, electrical current is cut to the electromagnetic door releases. Without the flow of electricity, the magnet loses strength. A separate door closer pulls the door shut. Fire codes specify where these emergency door releases are required.

INTERESTING FACTS
These magnets are amazingly strong. Try to close a door held open by one. You will have to tug to get the magnet to release.

Escalator

BEHAVIOR

Takes you to the top, or at least to the next floor. It conveys you up an incline so you can easily reach higher floors in a building.

HABITAT

You can find escalators in large office buildings and stores. Inefficient at moving people up more than one or two floors, they are great at moving many people up one floor at a time.

HOW IT WORKS

Escalators are moving stairways. You stand on a step and it carries you to the top. Each step rests on a conveyor belt that is pulled by motors.

Each step is like a railcar supported by its wheels that sit inside tracks. Unlike rail cars that run on two tracks, escalator wheels are set to run on four tracks, two on each side—one for the front wheel and one for the back wheel.

A powerful electric motor, typically 100 horsepower, at the top of the escalator drives a gear that pulls the steps upward. Descending steps counterbalance the ascending steps, reducing the load on the motor.

The motor also drives the handrail by a drive belt. Engineers try to design the system so the handrail keeps pace with the steps. Check to see if yours does.

UNIQUE CHARACTERISTICS

Whenever you see an escalator under repair, check it out. Often you can peer into the work space beneath the escalator to see its complex inner workings.

The grate at the top is designed to prevent shoes or feet from being caught in the machinery, but I can tell you from personal experience they don't always work. My son's shoe jammed in a grate and stopped an escalator on a Friday afternoon. It was still there Monday morning.

Elevator

BEHAVIOR

Gives quick vertical transportation to the floors in an office building.

HABITAT

Found in most multistory buildings.

HOW IT WORKS

There are two types of elevators: those for buildings with only a few floors and those for high-rise buildings. Smaller buildings typically employ hydraulic elevators. The elevator cab rides atop a piston housed in a shaft that descends into the ground beneath it. To rise, pressure is exerted on hydraulic fluid beneath the piston, which forces it to rise. You can often hear the pump start to operate when the cab moves. These systems are similar to the hydraulic lifts used in service stations.

In taller buildings, pulley systems are preferred over hydraulic systems. Pulleys and motors are mounted in a small shack on the roof. The weight of the cab is counterbalanced by a set of metal or concrete weights so the motor doesn't have to lift the entire weight of the cab and riders. The cab rides along vertical guide rails. In addition to the cable that lifts the cab, a second cable connects it to a safety break. Should the lift cable part and the cab start to fall, the safety break spins and pushes arms outward through centrifugal force, and arrests the fall.

Regardless of type, all elevators have a call system and a computer program for figuring where the cab should go to respond most efficiently to calls.

INTERESTING FACTS

Although known as the inventor of the elevator, Elisha Otis didn't invent the elevator; he invented the safety brake for elevators and then used it to create the safety elevator. Before his invention few buildings rose above five floors as people didn't want to walk higher and elevators without safety brakes weren't safe. His invention, along with the development of stronger steel for construction, changed the landscape of cities throughout the world, both horizontally and vertically.

Time Clock

BEHAVIOR

It stamps the date and time that employees arrive at work and leave work, creating a record of the hours they spend on the job.

HABITAT

Mounted on a wall, very close to the employee entry. Adjacent to the clock itself will be a vertical bin holding the time cards for each employee.

HOW IT WORKS

Employees coming to work pick their card from a bin hung on the wall adjacent to the clock. Inserting the card into the time clock (time recorder) trips a lever that stamps the date and time on the card. At the end of the day employees repeat this process to show how long they worked.

These mechanical systems are rapidly being replaced by more modern technology. Some systems use magnetic cards that identify the employee. On entering and leaving work, workers swipe the magnetic card under the reader to record their time.

INTERESTING FACTS

The first time clock was invented in 1888 by a jeweler. His brother started manufacturing time clocks, and this company was one of several that merged to form IBM.

Panasonic

ON THE OFFICE FLOOR

BESIDES THE CRUMBS of potato chips from yesterday's lunch and those annoying paper punch circles, what's on the floor of your office?

Most office technology is located above the floor, in and on desks, shelves, file cabinets, and stands. But some devices sit firmly on the carpet; these are featured in this chapter.

The floor is also home to boxes of records and things that have no other home. If you don't know where to put it, drop it in a box and slide it under a desk or into a closet. This, of course, is not a good policy as water pipes do break, flooding offices, and people do stumble onto stuff they shouldn't.

My three-year-old, not feeling well, spent a day with me in my office years ago. I left the room for a few minutes and on my return found that he had discovered a bottle of liquid that we used to tranquilize fish at the South Florida Aquarium. Without a good place to store this toxic chemical, I had shoved it under my desk thinking that it was a safe spot. We spent 30 frantic minutes on the telephone with the poison center without being able to identify any remedial action we should take, and realized that whatever might have happened wasn't going to. I got rid of that chemical, but I still have my financial records in a box jammed underneath the file cabinet of my desk.

Copier

BEHAVIOR

Reproduces documents, making one or several copies quickly. Some collate and staple documents as well.

HABITAT

In nearly every office, this machine dominates the common area or document preparation area. With its back to a wall and plugged into an electrical outlet, it is one of the most frequently used office machines.

HOW IT WORKS

When you place a document on the glass plate and press the start or print button, a bright light illuminates. The light reflects off the paper, but more light is reflected off the white areas than the dark areas (the printed words or figures). The reflected light shines on a special drum that is the heart of the process.

The drum starts out each cycle by receiving a strong positive charge of static electricity from the "corona wire." The surface of the drum is coated with a semiconductor that responds to light, so areas that are hit with the reflected light lose their positive charge. Now, the surface of the drum has an electrical image of the document with positively charged areas corresponding to letters printed on the original document and neutrally charged areas corresponding to white areas.

The drum rotates to pick up toner or dried ink. The toner, which has a negative charge, adheres to the drum in places that have positive charges, which represent the letters in the original document.

A piece of paper is fed from a hopper and the drum rolls over it. The bits of negatively charged toner adhere to the paper, which has been given a positive charge. The paper runs through a set of rollers to heat and compress the toner, fixing it in place. The paper comes out of the machine, warm and with a slight electrostatic charge.

INTERESTING FACTS

No one wanted the photocopier when it was first invented. Chester Carlson came up with the working model in 1938 after years of experiments. He presented his invention to nearly two dozen companies, but none saw the value in it. After all, what good is a photocopier when everyone had carbon paper (that made one to three copies as a secretary typed the original document)? Eventually a company that sold photo paper bought a license to manufacture the machines (1948), and after sales took off the company renamed itself Xerox (1959).

Paper Shredder

BEHAVIOR

Cuts paper documents and unwanted credit cards into strips for safe discarding.

HABITAT

Most shredders have their own dedicated trash bin into which the cut paper falls. They are located near electrical outlets in an out-of-the-way area of the office or near the copier and other document preparation machinery.

HOW IT WORKS

An optical sensor detects when paper is inserted into the shredder. The sensor turns on the motor that drives the blades. Some shredders also have manual on/off switches.

Paper passes between two parallel bars each supporting disks with sharp teeth. The disks interweave, leaving little space for the paper to slide through except between the teeth. The motor spins the bars and the toothed disks pull the paper in, chew it up, and spit it out into the bin below. Nefarious people can, with much trouble, reassemble documents shredded with this inexpensive shredder.

More expensive shredders cut both horizontally and vertically. This provides an even greater degree of security, but at a price. These shredders generally cost more and require more maintenance.

INTERESTING FACTS

You can "unshred" shredded paper. It's laborious, but can be done, like a very difficult jigsaw puzzle. So if you don't want anyone to be able to read the documents you're shredding, mix the shreds in different waste bins.

Adjustable Chair

BEHAVIOR

It swivels, it rises and lowers, it holds you in its grip for hours at a time, hopefully in a comfortable position.

HABITAT

Most work spaces have office chairs. Not in use, they are pushed into and partially under the desk. In use, most are parked in front of a computer. At the annual party, they can be rolled and raced down the corridor.

HOW IT WORKS

Office chairs are designed to improve efficiency. They allow the sitter to turn and access adjacent files and equipment without having to stand up. Many have casters to allow the user to move from side to side to reach even farther. Casters are wheels that align with the direction they are pushed.

Many chairs are vertically adjustable. Pushing on a lever disengages them from a locked position. An internal spring then pushes the chair upward. Sitting on the chair will push it downward. The downward movement is slowed by a pneumatic piston so that you don't hurt yourself if you accidentally kick the lever. Setting the lever again locks the chair at a selected height.

Water Cooler

BEHAVIOR

In addition to being a magnet for malcontents and talkers, the water cooler dispenses drinking water, often both hot and cold.

HABITAT

The cooler is often located in a kitchen area or near the coffeepot that many offices have.

HOW IT WORKS

Water arrives in five-gallon plastic containers, called "bubble tops." Since five gallons weighs 40 pounds, not everyone can easily pick up a bubble top, invert it, and place it into the cooler. So when the old bubble top runs dry, the call goes out to those of strong arms and backs to replace it.

From the bubble top, water flows into your cup when you open a spring mounted valve. The system allows air to bubble back into the plastic jug to prevent a partial vacuum. Many units have a chiller and heater to provide users with cool and hot water.

Another style of water cooler takes water directly from the municipal water supply and chills it. This type of cooler is mounted on the wall and has an electric cord that powers the internal refrigeration unit.

UNIQUE CHARACTERISTICS

Most people assume that the water in a cooler is spring water, but check out the bubble top. Some companies get their water from the municipal water supply.

INTERESTING FACTS

Older models required you to remove the plastic lid, leaving the bubble top open as you inverted it to place it on the cooler. Newer models have lids that don't require removal. Placing the bubble top upside down in the cooler punctures the lid (after the bubble top is seated in position), which starts the flow of water.

Air Filter or Purifier

BEHAVIOR

Removes airborne particulates; some models kill bacteria and viruses in the air.

HABITAT

Found in offices near workers who are concerned about their health.

HOW IT WORKS

Filters and purifiers remove allergens (particles that cause allergic reaction in some people), dust, and smoke from the air. Some filters have fans that push air through a filter medium to catch particulates. Their filters need to be replaced or cleaned periodically.

Devices that have high efficiency particulate absorbing (HEPA) filters are the most effective at removing even the smallest particles. They remove essentially all of the particles larger than 0.3 microns/micrometers. Because the filters are so dense, moving air through them requires a strong fan motor.

Charcoal filters are effective at trapping gases. Activated charcoal has tiny pores in which the gasses get trapped and adsorbed by the carbon atoms.

Other filters move air without a fan, instead using electrostatic charges to attract ions in the air. Negatively charged plates inside the filter attract and hold particles (most of which have positive charge), but also generate ozone (a molecular form of oxygen containing three oxygen atoms—O_3—that is considered a pollutant). So some models have a guard that converts any ozone generated back into harmless oxygen molecules, which have two atoms of oxygen (O_2).

Another feature of some air filters is an ultraviolet lamp that kills airborne viruses and bacteria. The lamps damage the DNA in the microbes and render them unable to reproduce.

ON THE WALLS

GOOD WALLS, good neighbors make. There are few things as emotionally charged in an office as walls. Who gets an office with real walls and who gets a cubicle? Who gets privacy? What do others hear over the top of partitions? People don't complain nearly as much about ceilings or floors (unless, of course, the ceiling is leaking).

Office walls not only contain voices and noises, they also keep the cold winds out and the employees in. More to the point of this book, they also provide surfaces for all manner of stuff and interesting technology to be affixed. Photos, calendars, and windows share the walls with fire extinguishers, clocks, cameras, and more. Electric, telephone, and data services usually are embedded in walls to keep them out of sight and protected from damage. In many offices, creative visions and wacky ideas are captured on whiteboards mounted on the walls.

Clock

BEHAVIOR

Keeps track of the time as it flies through the day. Provides a major point of focus for a certain percentage of workers.

HABITAT

High on the wall, usually in the center of the room.

HOW IT WORKS

There are several ways to keep time, and three are commonly used in offices. A traditional electric clock is powered by the 110V alternating current from a wall outlet. You can see the power cord connecting the clock to a wall outlet. The clock can also be wired directly, in which case you will not see a cord.

The clock uses a motor to convert electricity into the movement of the hands that show the time. The motor moves at a precise speed, regulated by the frequency of the electric current coming from the electrical generating plant.

A clock that doesn't have a power cord supplying it with electricity probably has a battery (often AA) inside. This type of clock uses a quartz oscillator to make consistent vibrations that are counted by an electronic circuit. The circuit controls a small motor that turns the hands.

Digital clocks use quartz oscillators too, but they count the time and convert it into a digital readout. The time is shown on a liquid crystal display (LCD) or as a series of light emitted diodes (LEDs).

A recent development is the use of the global positioning system (GPS) to keep clocks on time. An antenna collects time signals from satellites and passes them on to a computer and transmitter. This device directs all the clocks in a building by transmitting the time data by radio.

INTERESTING FACTS

The ancients defined an hour as one-twelfth of the time from sunrise to sunset. They divided the daylight into 12 segments, and night into 12 more segments, apparently to mimic the 12 lunar cycles of a year. Sunrise marked the first hour of the day, and sunset marked the end of the twelfth hour.

Over the centuries many rulers decreed changes to time systems. The modern standard time zones are a relatively new invention. In 1883 railroads in the United States and Canada established standard times for five zones in North America. Their time zones solved the problem of every city keeping its own time based on when the sun was at is zenith. Local time made town clocks very important, but railroad schedules impossible to understand.

Electric Outlet

BEHAVIOR

Allows you to connect appliances and devices that use electricity to the wires in the walls that carry electricity.

HABITAT

Outlets are everywhere, except of course, where you need them. Even just a few years ago builders didn't anticipate today's demand for electricity, so older offices have noticeably fewer outlets. Or, they have electrical conduit attached to walls providing additional outlets. Most outlets are located about a foot above the floor, mounted in a wall.

HOW IT WORKS

One slot in the outlet is "hot" and delivers 120 volts, while the other slot is neutral or at 0 volts. Electricity flows from the hot side, through the appliance, to the neutral side. The third hole in the outlet (this one is round), required in the United States since 1965, is "ground."

Appliance plugs have two flat metal pins that run parallel to each other and fit into the hot and neutral slots of outlets. The third post on a plug is a round metal pin, slightly longer than the other two. This round connector provides a "ground" connection, and its length ensures that a circuit or appliance is grounded before it is energized with current. Grounding ensures that should the hot wire somehow come in contact with the case of the appliance, anyone touching the case won't get shocked. Instead, the errant electric current will flow to the ground wire.

UNIQUE CHARACTERISTICS

Many appliances come with electrical plugs that have different sized pins. A three-pin plug has a round ground pin, but the other two pins may be different sizes as well. This ensures that you will insert the plug only in the desired way. It also keeps the hot side of the circuit connected to the switch inside the appliance, providing one more measure of safety.

Look at the ends of the flat metal pins of an electric cord. You'll see a small hole in each pin. When the plug is inserted into an outlet, two corresponding dimples inside the outlet fit into these holes to help hold the plug in place.

INTERESTING FACTS

Harvey Hubbell invented the electric outlet and plug in 1904. He also invented the pull-chain switch for electric lights.

A short circuit occurs when the current goes directly from the hot side to the neutral side without going through an appliance. With no resistance (the appliance) in the circuit, the current is quite high and can quickly heat wires to the point that they melt their insulation and cause fires. Circuit breakers detect very large currents and automatically disconnect the circuit before damage is done.

Telephone Jack (RJ-11 Connector)

BEHAVIOR

Provides a quick connection for a phone or modem.

HABITAT

Found throughout an office, usually above the baseboards, mounted in a wall.

HOW IT WORKS

A "jack" is a socket that the telephone plug fits into. The RJ stands for "registered jack," meaning that it is a standard connector. The jack connects the two center wires, usually red and green, coming from the switchboard or switch through the wall. You insert the plug into the jack and lift the receiver to get a dial tone. A small plastic tab locks the plug in place so it doesn't pull out easily. You have to press the tab to remove it. A jack that provides service for only one phone is an RJ-11. An RJ-14, also common, connects up to two phones from one jack.

INTERESTING FACTS

Bell Telephone Laboratories invented registered jacks in 1973 and installed them in phones across the country. When the phone monopoly was broken up in 1984, the Federal Communications Commission codified the use of registered jacks. In addition to the RJ-11, there are dozens of other types of registered jacks.

Power Supply

BEHAVIOR

Supplies a voltage other than 120V AC to a variety of devices, from printers to modems to answering machines.

HABITAT

Power supplies hang out at electrical outlets everywhere. When not plugged in, they're packed for travel or sit in a drawer.

HOW IT WORKS

Look at the front and back sides of the power supply. Printed on one side is what it expects for an input voltage—120V AC in the United States or 240V AC for many other countries—and what voltage it outputs. The output can be alternating current (AC), in which case the device is just a transformer. It changes the voltage from the line voltage (120V) to what the appliance needs to operate.

Inside a transformer are two coils of wire, each wrapped on opposite sides of an iron core. One set of windings connects to the line voltage, and the other connects to the wire that powers the appliance. The two sides will have a different number of windings around the core. The ratio between the numbers of windings determines the voltage output: fewer windings on the appliance side give lower (than line) voltage.

If the output is direct current (DC), the power supply transforms the voltage to the appropriate voltage and "rectifies" the current, meaning it changes it from alternating to direct current.

UNIQUE CHARACTERISTICS

Even when you're not charging your cell phone, if its power supply is plugged in, it is using energy. Touch a plugged-in wall wart—it will feel warm. The warmth tells you that energy (company dollars) is being wasted. Add up the pennies lost each day for each power supply, and over the course of a year the money wasted throughout an office could pay for an office party.

Switch

BEHAVIOR
Allows you to control the flow of electricity to lights and other appliances.

HABITAT
You find switches at chest height along the inside wall of rooms. You can tell the location as it is often dirty from errant fingers trying to find the switch in the dark.

HOW IT WORKS
The switch lever is made of plastic, which does not conduct electricity. On the inside of the switch, the other end of the lever pushes on a metal yoke that makes contact between two sets of terminals. A spring holds the switch lever in place in both the "on" and "off' position.

Some switches don't have the snap-action of a spring switch and instead are mercury switches. These noiseless switches rotate a drum of mercury that provides the path through which electricity can flow. In the "off" position, a nonconducting divider keeps the mercury from contacting both sets of terminals. Rotated to the "on" position, a hole in the divider allows the mercury on each side to connect.

INTERESTING FACTS
Over time you develop an innate understanding for where switches will be found. Traveling to different countries can leave you groping for the switches, as they are not where they're "supposed" to be. Sometimes they are located on walls outside the rooms they serve and sometimes at knee level rather than chest level.

Two-Way Switch

BEHAVIOR

Allows you to turn a light on or off from either end of the office. Think about that for a moment and then try to draw a diagram of the wiring that would allow the switches to work from two locations. Pretty tricky.

HABITAT

Two-way switches are mounted on walls in rooms that have two widely spaced entrances. You might want to turn on the light as you enter the room from either side, so a switch is available at each entrance.

HOW IT WORKS

One side of the light is connected to power, and the other side is connected to one of the switches. The two switches are wired together so that when they are both in the up position or both in the down position, they complete the circuit and the light comes on.

The switches themselves are different from ordinary electrical switches: they have an extra set of contacts so they can be connected. The first position—there is no set "on" or "off" position—in the first switch is wired to the first position in the second switch. And, the second position of each switch is connected.

One switch is connected to the light, and the other switch is connected to a wire providing electrical power. When both switches are in the first position, current flows into the first switch, across to the second switch, and out to the light. The same thing happens when both are in the second position, but the current is blocked when one switch is up and the other is down.

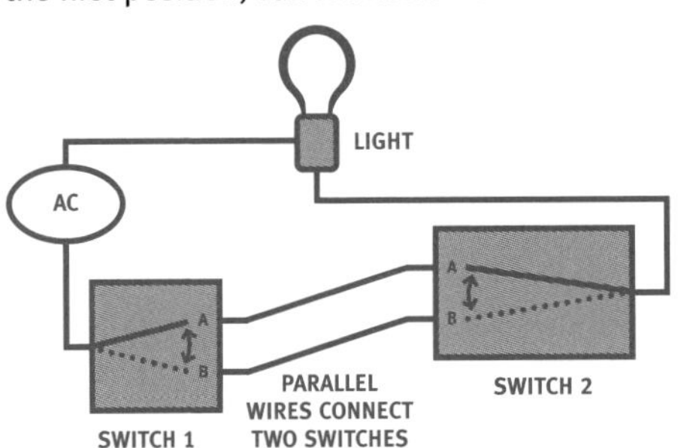

Circuit Panel

BEHAVIOR

Houses the circuit breakers and the electrical connections between the wires that bring electricity into your office from outside.

HABITAT

It is built into the wall, usually in an equipment closet or other out-of-the-way location.

HOW IT WORKS

The electric wires from the utility pole or underground contain three wires: two wires connect to opposite ends of an outside transformer (hung on the utility pole or mounted on the ground) and one wire comes from the center of the transformer (providing the neutral connection). Most devices in your office need to connect to one of the two wires tapped to the ends of the transformer that supply 120 volts. To complete the circuit, the appliance is also connected to the neutral wire so electricity flows from one "hot" side through the appliance and to the neutral side. A fourth wire, the ground, connects the circuit panel itself to the ground through a thick grounding strap made of woven wire.

Inside the panel, electricity is parceled out to different circuits. Open the panel door and read the labels. You are safe to open the door, but removing the panel cover (removing the screws) exposes wires carrying electricity. You might have one circuit for each room within an office or for each area of an office. Appliances that use lots of electricity—like air conditioners, electrical heaters, and large copy machines—typically have their own circuit breakers.

UNIQUE CHARACTERISTICS

If a circuit breaker "breaks," that is, if you lose electricity to part of your office, check first to find what tripped the breaker. Someone might have just overloaded the circuit by using one too many space heaters on a cold day. In this case, plug one of the heaters into a different circuit with an extension cord. To prevent people from tripping over the cord, get an electrician to run an additional circuit to that part of the office.

If adding an additional appliance didn't cause the circuit to break, look for a possible short circuit among the wires and plugs feeding appliances on the blown circuit. When you have removed the problem, find the blown breaker. You can usually see that it is out of position. You can also wiggle it; it will feel less secure than the other breakers. Pull the errant breaker to the "off" position and then toggle it back to the "on" position. If it trips again, you know there is still a problem in the circuit or appliance, a problem that requires professional attention.

Circuit Breaker

BEHAVIOR

Protects property and lives by automatically shutting off the flow of electricity if the flow becomes excessive.

HABITAT

Circuit breakers reside inside the electrical distribution panel (see Circuit Panel entry).

HOW IT WORKS

Short circuits can occur if two electric wires, one carrying electric energy and the other connected to the neutral or ground, come in contact with each other. For example, if the insulation is damaged on the two wires that power a lamp and the wires touch, the circuit is "shorted" and the breaker "trips."

Inside the breaker, excessive electrical current energizes an electromagnet that pulls a lever to shut the circuit. To reset the circuit breaker (after fixing the problem that caused the short) you merely open the circuit case, push the switch to its full "off" position, and return the switch to its "on" position.

One master circuit breaker protects all the other circuits. It is in the top center of the distribution panel. Each circuit in an office has its own circuit breaker. These are mounted in two columns inside the distribution panel.

Thermostat

BEHAVIOR

Allows you to set a desired temperature and cycle the heater or air conditioner on and off to maintain that temperature.

HABITAT

Thermostats are usually mounted on a hallway wall, about five feet above the floor. They are located where there is no direct source of heat or cooling, away from outside walls and windows.

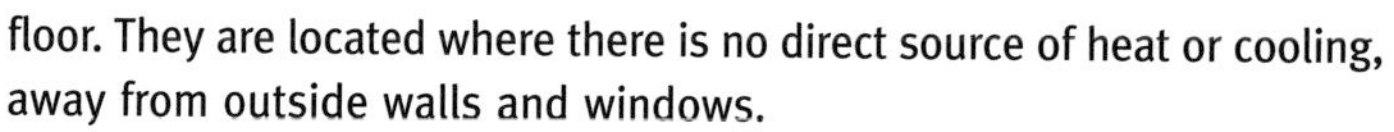

HOW IT WORKS

Older thermostats use bimetallic strips that bend as the temperature changes. Two strips of different metal are laminated to each other; when heated, the bimetallic strip bends. The two metals are chosen for their different rates of thermal expansion.

The strip is wound into a coil, spiraling from the center of the thermostat. At the far end of the strip is a glass vial holding a mercury switch. A tiny amount of mercury inside can close the contacts of a switch when the vial is positioned correctly. Electric current flowing through the switch closes a relay that starts the furnace or air conditioner. As the coiled strip moves as a result of temperature change, the position of the glass vial changes, the mercury flows away from the switch, and the furnace or air conditioner stops. This type of thermostat is common in homes and older offices.

Newer offices have digital thermostats, which use a different technology. They measure temperature using an electrical component called a "thermistor." It is a resistor whose resistance changes as temperature changes. The thermistor is part of a circuit that measures the temperature and switches the HVAC (heating, ventilating, and air conditioning) on and off. These newer thermostats display the room temperature with an LCD display, and some can be programmed to raise and lower the temperature at times throughout the day. Most use battery power.

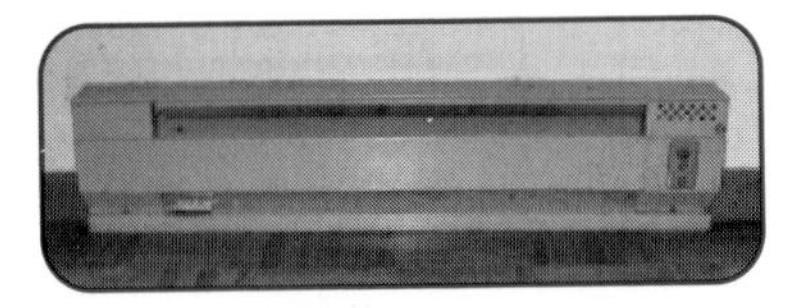

Electric Baseboard Heater

BEHAVIOR
Transforms electrical energy into heat by passing electric current through high-resistance wire to heat air.

HABITAT
Along walls at floor level. Baseboard heaters are more common in smaller offices in parts of the country that enjoy lower electrical energy rates.

HOW IT WORKS
Electricity moving through metal wires generates heat from the movement of electrons. Some metals, like copper and silver, have very low resistances and generate little heat when electricity passes through. Other metals, most notably nickel-chromium compounds, are highly resistive. Such metals are used in toasters and heaters. Nickel-chromium contains about 60 percent nickel, 16 percent chromium, and 24 percent iron.

As current flows through the wires they get warmer and heat the adjacent air. Warm air rises and cooler air near the floor replaces it, setting up convection currents that heat the room.

UNIQUE CHARACTERISTICS
Baseboard heaters sport fins that transfer the heat from the heating element to the air. Fins provide a large surface area so more air molecules come in contact with the hot surface.

INTERESTING FACTS
Baseboard heaters have a thermal shut-off switch. If the heater becomes dangerously hot (possibly because the flow of air to the heat is blocked), the liquid inside a switch expands and presses against a diaphragm to open the switch. If the switch opens, you may have to reset it by pressing the reset button where the electric wires enter the heater.

Radiator

BEHAVIOR

Heats the office by transferring heat from hot water inside to the surrounding air.

HABITAT

More common in older offices, radiators are attached to a wall or floor.

HOW IT WORKS

The radiator is the delivery vehicle for heat. Hot water or steam is forced through the radiator. Fins mounted on the hot water pipe inside the radiator increase the surface area so more air can come in contact with the heat. Convection currents (warm air rising, cold air sinking) carry away warm air and bring in a supply of cooler air to be heated.

Pipes leading to the radiator carry water that has been heated in a hot water boiler to 180 to 240 degrees Fahrenheit. Valves throughout the system allow for individual radiators to be cut off from the flow of hot water if a room doesn't need to be heated. Some systems have thermostats in each room that control the valves to adjust the temperature. Return pipes carry the water that has passed through radiators back to the boiler for reheating.

Hot water radiator systems are good at maintaining a steady temperature in a large building but are slow to respond if you want to change the temperature.

INTERESTING FACTS

Although the Romans circulated warm water through houses for heating, central heating did not appear again until the eighteenth century in France.

Pneumatic Tube/Vacuum Tube

BEHAVIOR

Whisks containers carrying important documents and parts to distant areas of an office or factory. Most often seen at drive-up tellers at banks, it transfers deposits and withdrawals between banker and driver without either leaving their seats.

HABITAT

You will find these tubes at drive-up banks, hospitals, manufacturing plants, and retail stores. Casinos, auto dealers, stockbrokerages, and warehouses all use pneumatic tubes.

HOW IT WORKS

A motor spins a fan that creates positive air pressure to move the "carriers" through the tubes. At a bank, after putting your deposit in the carrier and putting the carrier in the tube, you push a button that closes and seals the door. That activates the blower or valve that lets a blast of air through the tube, driving the carrier forward.

Most systems are point to point, which means that the system transports carriers between two points. The carrier moves in one direction, going to the teller in one set of pipes and returning to the driver through a second set of pipes. However, hospitals and other users have multiple station systems where a sender can direct the carrier to one of several destinations.

INTERESTING FACTS

Pneumatic tube systems were the rage in the nineteenth century. Some cities delivered mail to offices and residents via tubes. Only one such system remains in operation today, in Prague, Czech Republic.

Whiteboard

BEHAVIOR
Captures ideas in a group setting and invites doodling. (If Kilroy was anywhere, he certainly was here.)

HABITAT
In offices of creative people, mounted on walls.

HOW IT WORKS
The whiteboard and dry erase markers replaced blackboards and chalk, which were used until the 1990s. The writing surface of a whiteboard is either plastic or enameled steel. Users draw and write with dry erase markers, which have nonpermanent ink. The ease of writing and erasing facilitates idea generation. Whiteboards are also used for grocery shopping and to-do lists.

UNIQUE CHARACTERISTICS
Permanent markers look almost exactly like dry erase markers, so people often mistake one for the other. Permanent marker trails can be washed off a whiteboard, but it is a laborious rubbing task.

Motion Detector

BEHAVIOR

Senses someone or something moving around in the office after the alarms have been set. If someone has managed to break into the office without setting off a perimeter alarm, a motion detector will sense him and signal an alarm.

HABITAT

You can spot these devices hanging high on a wall, often in the corner of a room. They are usually the same color as the wall, so they blend in. They have what appears to be a curved lens, and some have an active light emitting diode (LED) to indicate that the sensor is working.

HOW IT WORKS

Most motion detectors use infrared imaging. All objects emit infrared radiation; the sensor detects major changes in the amount of infrared radiation in the room. It sends an alarm when it detects a significant change in the infrared energy. Someone entering an empty room will add infrared radiation that the detector can distinguish.

The combination of perimeter and motion detectors makes it quite difficult for anyone to enter undetected. Unless someone wants to go to the lengths of *Mission: Impossible* to get in, it's easier to find a less protected office.

Security Camera

BEHAVIOR

Allows security guards to view areas inside and outside an office so they can alert police or fire departments if needed.

HABITAT

Cameras are usually installed at the corners of a building, near elevators, in the lobby, and in other areas of egress. The monitoring station where the camera's video is viewed is at the guard station in the building or at an off-site security company location.

HOW IT WORKS

Most security cameras are fixed in position, but some may be moved for better viewing by the observing guards. Unlike broadcast television, the signals are carried by wires from the camera to the monitor and any recording equipment. Security cameras are a type of closed-circuit television.

Cameras are usually mounted high enough to prevent tampering. Light enters through the lens and is detected by a charged coupled device (CCD). The CCD has an array of tiny capacitors (electronics components that hold charges). Light reflected off objects in front of the camera is focused on capacitors on the CCD. Each capacitor holds a charge in proportion to the amount of light that hits it. Thus the image projected by the lens is transformed into an array of electric charges. This array is read by an electronics circuit several times each second and sent to a monitor for viewing.

Monitoring is done in real-time, but the video is also taped and kept until it is determined not to be needed.

Emergency Lighting

BEHAVIOR

Illuminates the office when electrical power is interrupted.

HABITAT

Emergency lights are mounted high on walls, aimed to provide illumination for hallways and exits, and secondarily for general office areas.

HOW IT WORKS

Incandescent floodlights or focused light emitting diodes (LEDs) are housed along with a battery in a case mounted on the wall. The battery, either 6 or 12 volts, is a lead or lead-cadmium battery that can last for many years. The battery is continuously recharged through a transformer powered by the office electric circuit.

When the emergency lighting system detects a voltage drop, it switches on its lights, which draw electrical energy from the battery. When power has been restored, the emergency light system has to be reset manually to the “off” position.

Lighted Exit Sign

BEHAVIOR

Provides directional egress information in emergencies. It shows you the way to go home . . . or at least the way to get out of the office.

HABITAT

These signs are located on walls above the doors that take you safely out of the building. They may also be located lower on the walls, pointing to such doors.

HOW IT WORKS

The light generator inside can be an incandescent or fluorescent light, or light emitting diode (LED). The lights draw power from a battery or uninterruptible power supply, which are both charged from the office electric power circuit.

Exit signs that use fluorescent bulbs (to save energy when electrical power is working) use incandescent bulbs when the power is lost. Incandescents (and LEDs) can run on lower voltage power (6 to 12 volts) that requires smaller and lighter batteries than fluorescent lights.

Some exit signs do not have electrical lights. These self-luminous signs do not need to be powered. Some reradiate the energy they absorb from ambient lighting, a process called phosphorescence. Others convert the radioactive decay of a material into light. This method is also used in some watches so you can read them at night.

UNIQUE CHARACTERISTICS

What color is that exit sign? Different countries have their own building codes that specify the color of exit signs. In the United States, exit signs are either red or green.

INTERESTING FACTS

The word *exit* comes from Latin: *ex* means "out" and *it* is a form of the verb "to go."

Fire Alarm Pull Station

BEHAVIOR

The alarm system detects fires and warns workers. This device is a manual switch for the alarm.

HABITAT

Sensors for alarms are often mounted in the ceilings along hallways. Human-activated alarms are mounted on walls.

HOW IT WORKS

Alarms can be tripped manually by pulling a lever, or automatically by one of several types of smoke sensors. Only building personnel can reset manual levers once they've been pulled. Leaving the levers in the "pulled" position shows fire teams which pull station was activated. The photo shows a manual fire alarm.

Sensors can detect smoke, heat, or the flow of water through an activated sprinkler head. If a fire sprinkler detects heat and turns on, the flow of water through the pipes can alert the alarm system that a fire is present.

There are two types of smoke detectors: optical detectors and ionization detectors. Optical detectors have a small source of light, an infrared LED, and a photodiode that detects light. Smoke entering the detector scatters the light either into the photodiode or away from it (a variation of the first design), which sets off the alarm. Ionization detectors sense the decrease in ionization (creation of charged particles between two electrodes) due to the presence of smoke.

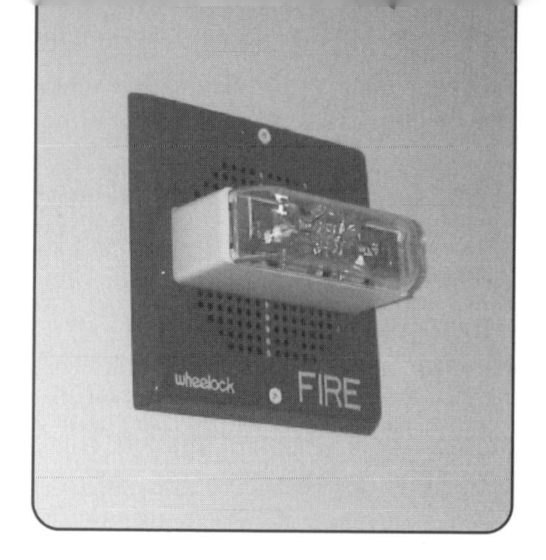

Fire Alarm Notification Appliance

BEHAVIOR

Flashes strobe lights and blasts out a noise so horrible you have to leave the area, even if you know it's a false alarm.

HABITAT

These devices are mounted high on walls where they can be seen and heard, often near the exit to help indicate the evacuation route.

HOW IT WORKS

Alarms can be reported to the control panel from an array of possible sensors including human-activated pull stations, heat or smoke detectors, or water pressure sensors in fire sprinkler systems. Once a report is received, the control panel engages all the notification appliances (commonly called fire alarms).

Alarms can be programmed to play different sound patterns, but all play at 70 to 90 decibels, loud enough to cause hearing impairment with long exposure.

Fire Extinguisher

BEHAVIOR

It quickly puts out the flames of a small fire. Because it has limited capacity, it should not be used to fight a raging fire.

HABITAT

Fire extinguishers are mounted on walls, sometimes inside a case with a protective door.

HOW IT WORKS

Most fire extinguishers put out fires by covering the burning material and excluding oxygen (from the air). Without oxygen, the fire goes out.

Inside the extinguisher are the smothering agent and a pressurized container with carbon dioxide. Squeezing the handle pushes down on a rod that punctures the carbon dioxide container. The gas is released into the interior of the extinguisher. Squeezing the handle also opens a valve that lets the now pressurized agent to escape through the nozzle.

The smothering agent in most extinguishers is a dry powder made of sodium bicarbonate, potassium bicarbonate, or monoammonium phosphate. Other extinguishers use carbon dioxide itself as the agent.

Fire extinguishers contain only enough material to last a few tens of seconds. They are suitable only for small fires. Larger fires require the response of a fire department.

UNIQUE CHARACTERISTICS

Check the pressure gauge on your fire extinguishers. With time, pressure is lost, making the extinguisher inoperable. Some can be recharged, but many of the lower cost extinguishers must be replaced when they have lost pressure. Most offices hire a service to check and maintain their fire extinguishers.

Defibrillator

BEHAVIOR

Used in medical emergencies, specifically heart attacks, to get the victim's heart beating regularly again.

HABITAT

Defibrillators are becoming more common in large offices and public buildings. They are mounted on walls in protective cases.

HOW IT WORKS

A defibrillator is designed to start the heart beating in normal rhythm. It is used when someone suffers ventricular fibrillation, which is the ventricular muscle twitching randomly rather than beating. The defibrillator supplies an electric shock that jump-starts the heart's control mechanism so it can again supply regular signals for the ventricular muscles to contract.

The person operating the defibrillator uses two paddle-mounted electrodes to deliver the shock(s). The paddles are placed on opposite sides of the chest. They have well insulated handles so the operator doesn't get shocked.

Defibrillators work somewhat like strobe lights in cameras. A battery charges a capacitor that delivers the shock very quickly. Capacitors can discharge much faster than batteries and so are ideal for supplying sudden bursts of direct current electric energy.

INTERESTING FACTS

Defibrillation was discovered in the late nineteenth century but not successfully used on a person until 1947. Portable defibrillators were developed in 1966 but have begun to be placed in offices only recently.

ON THE CEILING

THE ODD THING ABOUT CEILINGS is that they are in plain sight, but people rarely see them. Or, they rarely notice them. What kind of ceiling does your office have?

Dropped or suspended ceilings are popular today. They are easy to install: workers hang metal wire from supports above and use them to hold a grid of lightweight metal rails. The ceiling tiles are held between the rails.

If you lift one of the tiles up you will see the messy side of technology. Amid the dust bunnies are pipes and wires running across the office, as well as ducting for heating, ventilation, and air conditioning (HVAC). Light fixtures either poke through the tiles or consume an entire panel, supported by the grid.

Many offices opt not to have the low ceilings created by dropped ceilings. In these offices, ducts, pipes, and wires run across the office in plain sight. Some are labeled, which makes it easy to figure out what they do and where they go. Often the ceiling and upper walls are painted darker to conceal all the messiness of the building's infrastructure. But in some offices, this messiness is celebrated with a coating of bright paint.

Whatever type of ceiling your office has, look up to see what technology resides there. See if you can find the entries in this chapter.

Incandescent Lightbulb

BEHAVIOR

Passing electric current through these ingenious devices transforms electrical energy into light (and heat).

HABITAT

In flashlights, lamps, light fixtures—hopefully everywhere you need to see in the dark.

HOW IT WORKS

Electricity passing through the filament of a lightbulb causes its temperature to rise to about 4,500 degrees Fahrenheit. The electrical current has to do work to overcome the resistance in the wire, and this becomes heat. Once at this elevated temperature, the wire in the bulb glows.

Without the glass protective cover, the filament would quickly oxidize (the metal would chemically combine with oxygen) and break. Inert gases (nitrogen, argon, or krypton) are pumped into the glass during manufacturing to keep air (especially oxygen) out.

The heat given off by a glowing bulb suggests that more energy is given off as heat than light. In fact, an incandescent bulb converts only 10 percent of the electrical energy into useable light; the rest is wasted.

INTERESTING FACTS

In the United States we credit Thomas Edison with inventing the lightbulb. Elsewhere in the world credit is often given to both Edison and Englishman Sir Joseph Swan, who independently created a working lightbulb at about the same time as Edison. In reality, neither invented the lightbulb; they made it work well enough to become a useful product. And Edison developed the electrical system, from generating stations to distribution wires to lamps, that made the lightbulb useable. The development of the electric lightbulb is not the "Edison did it" tale, but instead a convoluted story of innovation, science, and legal battles.

Fluorescent Lightbulb

BEHAVIOR

Produces light at a lower cost with less heat generated (and less energy wasted) than an incandescent bulb.

HABITAT

Fluorescent lights are used in offices, mounted on ceilings. Often light fixtures are mounted in the grid of dropped ceilings.

HOW IT WORKS

The most interesting aspect of a fluorescent bulb is that the light generated by the gas inside the tube is invisible. Plug it in and you're still in the dark! However, the inside of the tube is coated with phosphors, materials that absorb light and then reradiate it. The light generated by the gas is in the ultraviolet range (hence invisible), but it excites the phosphors to give off light in the visible range. So, it is the coating on the inside of the glass tube that provides visible light.

When the light is turned on, the starter heats up and emits electrons into the gas tube. The fast-traveling electrons collide with gas atoms and cause their electrons to take on more energy. The higher energy state is unstable, so the electrons quickly give up energy in the form of light. The gases used in fluorescent bulbs give off light in the ultraviolet range.

Fluorescent lights are about three to four times more efficient than incandescent lights and last several times as long. Their higher cost is generally offset by their energy efficiency and longevity.

UNIQUE CHARACTERISTICS

Can you see the light from a fluorescent tube flicker? It turns on and off 120 times each second, much too fast to really see, but you might notice a visual blur. You won't see that with an incandescent bulb, although both types use 60 cycles per second alternating current. The incandescent bulb filament doesn't cool fast enough to stop emitting light. That is, it has thermal inertia that keeps the light steady.

INTERESTING FACTS

The hum you sometimes hear from a fluorescent light is usually the ballast. The ballast is what starts the lighting operation by using line voltage to send electrons flying through the gas. Since the line or house voltage alternates at 60 cycles per second (in the United States), it goes through zero volts twice a second (once rising to 120 volts and once falling to -120 volts). Thus the hum you hear is 120 cycles per second, close to the note B2. Although you might not detect the flicker of the light going on and off 120 times each second, some people report that it gives them headaches.

> The modern version of the fluorescent bulb was invented by Edmund Germer in 1926.

Compact Fluorescent Lightbulb

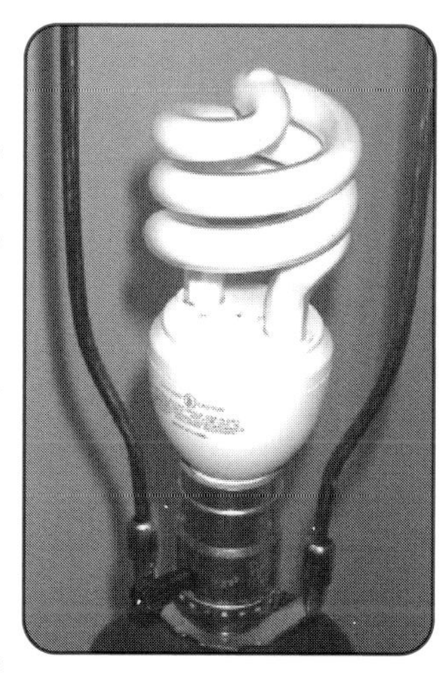

BEHAVIOR

Screws into the socket for an incandescent bulb, but operates as a fluorescent. It provides more light for less energy cost than an incandescent.

HABITAT

In discriminating light fixtures everywhere. Compact fluorescent bulbs are becoming more popular.

HOW IT WORKS

Compact fluorescent lamps are fluorescent bulbs with a built-in ballast or starter. As in traditional fluorescent lights, the ballast shoots out electrons into the gas-filled tube, adding energy to the gas atoms. As the atoms return to their normal state, they give off the excess energy as light. The light generated is ultraviolet—great for a sunburn, but otherwise invisible. To make visible light, the inside of the glass is coated with phosphors that absorb the ultraviolet light and reradiate light at visible wavelengths.

INTERESTING FACTS

An incandescent light is only about 10 percent efficient; the other 90 percent of the energy consumed goes to generating heat or light outside the visible range. Fluorescent bulbs are about three to four times as efficient and can last 10 times longer. Although compact fluorescents cost more to purchase, they save money in the long run.

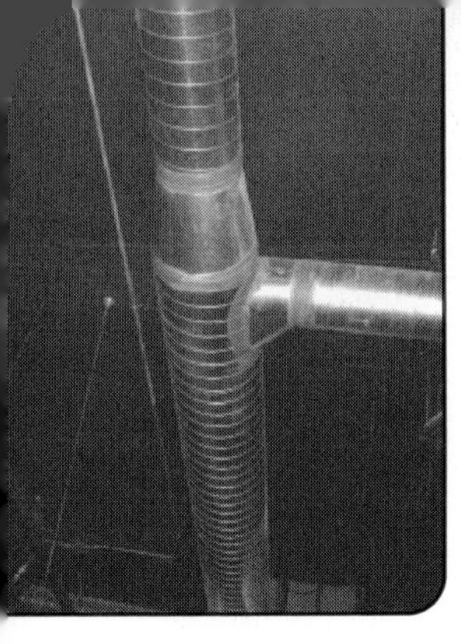

Heating Duct

BEHAVIOR

Carries hot (and cold) air in a central heating or HVAC (heating, ventilating, and air conditioning) system. In most buildings the ducts are hidden from view behind walls. However, it has become fashionable to hang the ducts from the ceiling. Ducts hung from the ceiling are usually round and painted, as opposed to the rectangular, unpainted ducts that are installed inside walls. This doesn't save much money on construction as the round ducts are more expensive, but it gives an office a modern look.

HABITAT

Most are the shiny, metallic, rectangular pipes found above ceiling tiles and behind walls. Large, round painted ducts are hung from the ceiling and exposed to view.

HOW IT WORKS

Locate the furnace or blower that supplies the ducts with warm air. Notice the size of the ducts. Near the furnace they are large. They narrow with each branch to a different office or different part of a large office. The branches lead to heat registers where the air is released into a room.

Ducts also return air from the office to the air handling machines (heater or air conditioner). Vents for return ducts are quite a bit larger than heat registers so they are less likely to be blocked.

INTERESTING FACTS

Duct tape was designed not for use on heating ducts but as a waterproof sealer for ammunition boxes during World War II. After the war people started using this versatile tape for a variety of uses, including making repairs on heating ducts. This is how the tape picked up its name. Today, most states discourage the use of duct tape on heating ducts. Professional repair people don't use it but may use a heavier, metallic tape. And, just to be confusing, they continue to call it duct tape.

Security Camera

BEHAVIOR
Provides wide-angle visual monitoring within an office or store.

HABITAT
Mounted on the underside of ceilings, housed in dark plastic hemispheres.

HOW IT WORKS
Behind the smoked plastic dome sits a video camera with a wide angle lens. Some cameras can be rotated and some can be tilted, rotated, and zoomed in and out by an operator. The dome not only protects the camera, it hides the direction in which the camera is looking. Most security cameras are not actively managed or viewed; they deliver video to a recorder that is examined only if needed.

The cameras operate in a closed circuit system—you won't be seen tonight on a new TV reality show. In low-light areas infrared cameras are used instead of cameras operating in the visual wavelength range.

INTERESTING FACTS
Some companies install fake domes—domes without a camera inside. If cameras act as a deterrent, then fake systems should work as well without the expense of installing and maintaining the video system.

Smoke Detector

BEHAVIOR

Detects smoke and alerts you to the possibility of a fire. Some are sensitive enough to detect a cigarette smoker lighting up inside.

HABITAT

At least one smoke detector should be mounted on the ceiling of each room throughout an office. Detectors should be at least six feet away from any walls. If mounted on a wall, the detector should be at least six inches away from the ceiling. Detectors are also required near elevator doors. Fire codes regulate the placement of these detectors.

HOW IT WORKS

Although some smoke detectors use a light or laser (smoke interrupts the beam of light, setting off the alarm), most use an ionization detector. The detector holds a tiny piece of the radioactive element americium. The americium radiation ionizes gases in the air—predominately nitrogen and oxygen—creating positive and negative ions. The ions permit a small current to flow between two charged plates. Smoke particles attach to the ions and reduce the current flow between the two plates. Sensing a decreased current, the alarm switches on that horrible buzzer.

Although home smoke detectors are typically powered by a 9-volt battery (don't forget to replace your battery), detectors used in offices are powered by the 110V power. Power is supplied to the central fire alarm system, which transforms it to the needed voltage for the detectors and converts it to direct current. The central system has a backup battery to power all the detectors and monitors in case of power failure.

UNIQUE CHARACTERISTICS

Some smoke detectors also detect heat or a rapid rise in the temperature. Separate heat detectors are installed in some places, but combination heat/smoke units are more common.

Fire Sprinkler

BEHAVIOR

Automatically releases water when heat rises above a threshold level.

HABITAT

They are supported either from the ceiling or from a wall in many offices and public buildings.

HOW IT WORKS

Each sprinkler has its own heat detector. When temperatures rise above the threshold, a metal link melts and the sprinkler releases water in the pipes. Other types of detectors use a chemical reaction or the breaking of a small glass capsule to trigger the flow of water. A pump or series of pumps responds to the drop in water pressure in the piping and turns on to maintain the water pressure.

Sprinkler systems can be dry or wet. Wet systems have water in the pipes that is under pressure; as soon as the sprinkler head opens, water comes out. In dry systems, the pipes are empty until a fire has been detected.

INTERESTING FACTS

Sprinklers have been used in the United States since 1874. They are required in new buildings that are taller than 75 feet, in public buildings where 100 or more people may gather, and in buildings where people sleep.

> How wet will you get standing under an activated sprinkler? Very wet. A sprinkler shoots out about 14 gallons of water per minute. Your shower at home sprays you with about two gallons per minute.

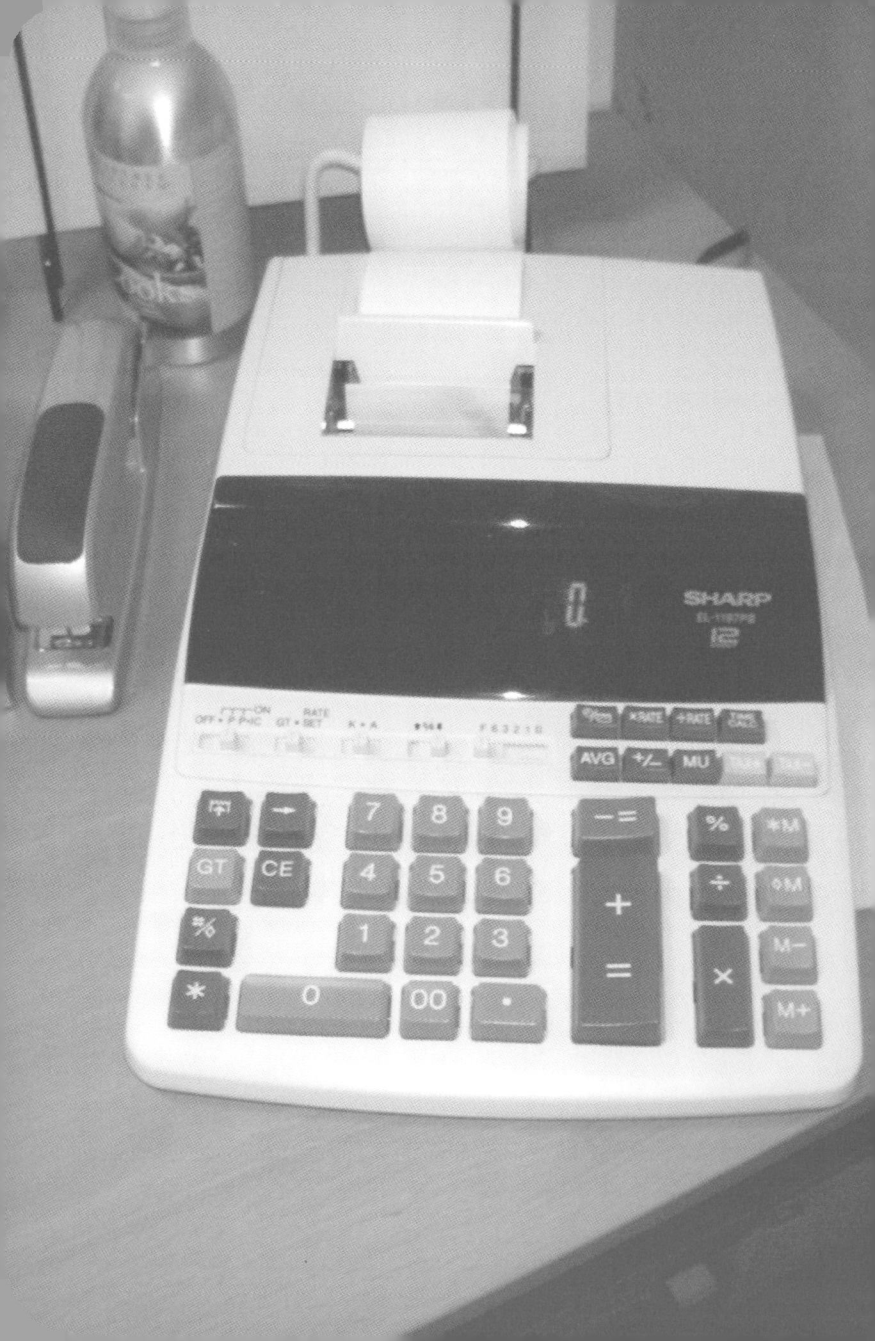
SHARP
AVG
MU
GT
CE
7
8
9
4
5
6
1
2
3
0
00
+
=
%
M−
M+

IN AND ON DESKS AND TABLES

OH, HOW WE ADMIRE a neat desk. Everything is in its place, with papers in well defined piles, and tools and materials not currently employed stowed away, out of sight. Is that how your desk looks?

Not mine! Stuff is everywhere. I leave piles of stuff so I won't forget the papers. Of course, I do forget them when other stuff hides them from view or when I've been so acclimated to seeing that pile that I don't notice it.

Personal work space—the desk and its immediate surroundings—says a lot about who we are and how we work. Even the messiest of us have some sort of system for storing and finding things in and on our desks. I want everything handy; the result of that impossible imperative is that often nothing is handy because everything is buried beneath all the other things I might need someday. If you need it, I have it, even if I can't put my finger on it right now.

For me, horizontal surfaces, like desks and tables, are for putting stuff on. If I put it in a drawer, I might not find it. And, I'm not quite sure which drawer to put it in. As the debris level on my desk and table rise, I recruit the other horizontal surfaces to augment them. Now, the floor has piles. When finally I can't navigate to my chair, I tidy up. "There, that wasn't so bad," I say. And soon I flop another pile of stuff on the floor.

This chapter breaks down the technology of the multitude of things in and on your desk that enhance—or clutter—your work space.

Pencil

BEHAVIOR

Allows us to record great ideas on the backs of napkins, to sketch designs, to solve Sudoku puzzles, and to make changes to all this great thinking with the handy eraser.

HABITAT

Usually stored either in the top wide drawer of a desk in a tray designed for them, or in a cup that sits on the desk.

HOW IT WORKS

Filled with a composite material (mostly graphite—not a drop of lead) and housed in a wood hexagon or cylinder, the pencil needs only to be sharpened to work. Sharpening cuts back the wood to expose the graphite and grinds the graphite to a point.

Pencil leads are made by combining ground graphite, clay, and water. This mixture is placed in molds and heated. The rods of lead are inserted into slots cut into a slat of cedar wood. An identical piece of wood with slots is glued on top. Individual pencils are cut from the wood, and the pencil is shaped for easier gripping.

INTERESTING FACTS

Although graphite has been used for marking and recording since the 1500s, the modern pencil with wood case was first made 200 years later. The eraser was added in 1858.

> How many pencils are made in a year? Take a guess. Laid end to end, the pencils produced in one year would encircle the earth more than 60 times. That's about 14 billion pencils.

Eraser

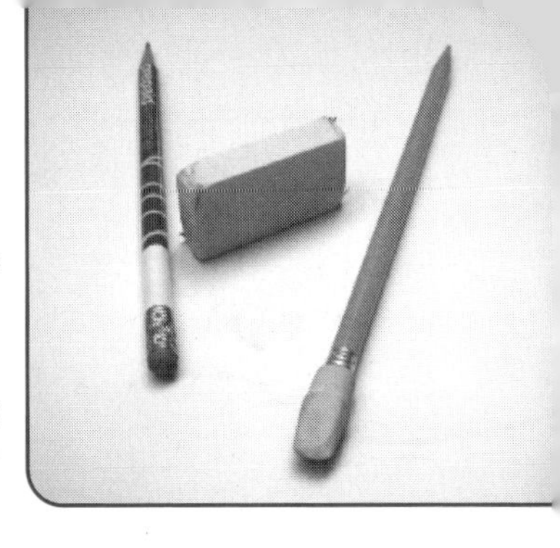

BEHAVIOR

Removes mistakes created with pencils, or even pens.

HABITAT

Perched nobly atop pencils, or in the top desk drawer.

HOW IT WORKS

Make a mistake with a pencil and quickly twirl it around to employ the eraser at the opposite end. Making the mistake left a trail of graphite (not lead) on the paper. Rubbing the eraser erodes the graphite, which sticks to the chunks of rubber along with tiny bits of the paper.

The rubber in an eraser is weakly held together, so pieces (carrying graphite and paper bits) fall off while the eraser is rubbed on paper. If you rub too hard you can erode your way through the paper leaving a gaping hole. The residue of this correction process is lumps of rubber and graphite, and bits of paper.

Other, larger erasers don't break apart as you use them. They collect the bits of graphite and eventually turn black. You bend them to get a fresh surface to use for future erasures.

INTERESTING FACTS

Edward Nairne invented the rubber eraser in 1770. It was one of his least complex inventions. Nairne's marine barometer accompanied James Cook on his second voyage of exploration.

Rubber erasers really became popular after Charles Goodyear discovered the vulcanization of rubber in 1838. Vulcanization keeps rubber soft and supple, and prevents it from disintegrating.

> The U.S. Supreme Court ruled in 1975 that adding an eraser to the end of the pencil was an obvious combination of two existing technologies and not worthy of a patent. This ruling invalidated the patent held by Hymen Lipman.

Pencil Sharpener

BEHAVIOR
With the flick of your wrist or a touch of a button its rasps spin to cut away the wood of a dull pencil.

HABITAT
Sits on a desk or on a cabinet or wall, waiting for pencils to dull.

HOW IT WORKS
Small mechanical sharpeners have a cutting edge set along the side of a conical opening. The pencil is pushed into the opening and twisted against the cutting edge.

Crank operated sharpeners have sharp burrs inside that cut away wood and lead to make a point. Shavings fall into a drawer that is never emptied until it overflows.

The Climax Pencil Sharpener, the predominant crank model used since 1910, has a cylindrical cutter with spiral cutting edges. Turn the handle and the cutters spin and rotate around the pencil.

Sharpeners today often have an electric motor that spins the cutting burrs around the pencil. When you insert a pencil it closes a switch and turns on the sharpener.

Pen

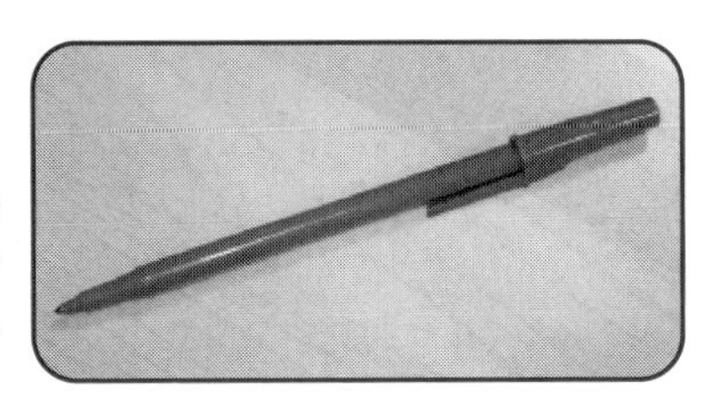

BEHAVIOR

It captures the wildly creative idea or the crucial signature by leaving a trace of ink on paper.

HABITAT

Pens are found in pocket protectors everywhere—and in unprotected pockets or the top drawers of desks.

HOW IT WORKS

An ink pen carries a reservoir of oil-based ink and a tip that dispenses the ink. Ballpoint pens are the most common. These have a tiny ball (1 mm diameter, or smaller) in the end. As the writer pushes the pen across paper, the pressure on the ball causes the viscous ink to flow. As the ball spins it carries ink to the paper, where the ink dries immediately. Inexpensive pens are disposable and more expensive ones have replaceable ink cartridges and balls.

Other types of pens have felt ink dispensers or use roller balls and water-based inks. Roller ball pens leave a wider and deeper trace on paper. Felt-tip pens wick the ink from the internal reservoir to the tip where it rubs off on the paper.

INTERESTING FACTS

Hungarian Laszlo Biro invented the ballpoint pen in 1938. Prior to his invention, people used fountain pens that required frequent refilling with ink. Biro grew frustrated with the time wasted in refilling the ink. Although Biro was awarded two U.S. patents, they weren't written well enough to prevent Milton Reynolds from becoming the first to produce ballpoint pens in the United States.

> The first writing instruments were probably sharpened sticks used to press letters in moist clay blocks around 3000 B.C. You have to wonder if they were carried about in pocket protectors.

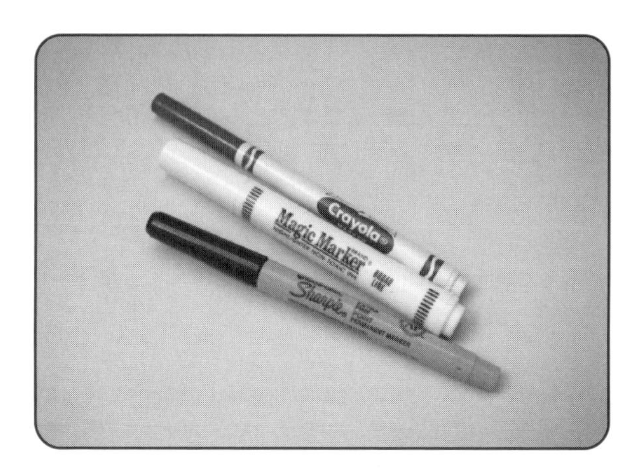

Marker

BEHAVIOR

It is used to leave words and images on a wide variety of surfaces.

HABITAT

When not in use, it usually resides in the top drawer of a desk.

HOW IT WORKS

Markers have a supply of ink in a cylinder and have a writing tip made of felt or nylon. As the tip is dragged across paper or other surfaces, ink flows onto the surface and more ink is wicked into the tip.

Most permanent markers have an alcohol-based ink; nonpermanent markers have water-based ink. Permanent markers' ink contains one of two toxic, non-water-soluble chemicals. Removing the cap from one of these markers releases some of the volatile chemical. The chemical can be absorbed through the skin.

Rubber Band

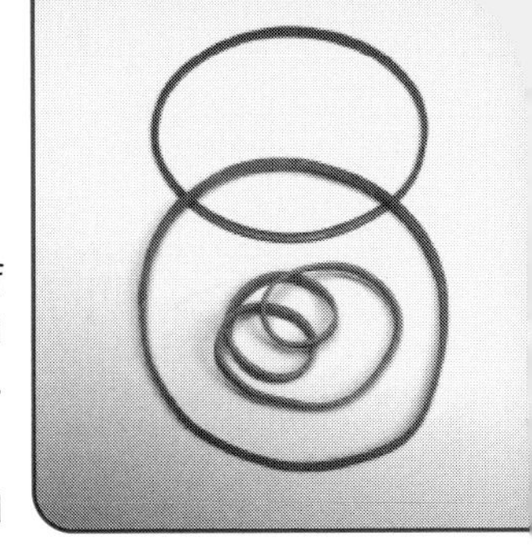

BEHAVIOR

It flies across the office from the fingertip of a bored employee, and it binds papers and other materials together with its elastic force.

HABITAT

Scattered throughout the desk drawers and across the desktops of messy employees.

HOW IT WORKS

The value of rubber bands is their elasticity. They stretch and rebound. The long polymer molecules in the rubber and latex of rubber bands are linked to their neighbors at distant intervals. Upon stretching, the molecules uncoil and stretch out, more or less parallel to their neighbors. Relaxing the stretch allows the molecules to coil up again. Exceeding the elastic limit breaks the bonds between molecules, causing the rubber band to break.

UNIQUE CHARACTERISTICS

Check out the temperature effects of rubber bands. Grab a big, fat one and hold it up to your forehead. Note the temperature. Now stretch the rubber band and hold it up to your forehead. You can feel the higher temperature. Pulling on the rubber band causes its molecules to pull tighter on their neighbors. As you pull on the rubber band, you are making the molecules work (to oppose your pull), and you sense that work as heat.

INTERESTING FACTS

Stephen Perry invented the rubber band in 1845, one year after Charles Goodyear invented the vulcanization process that made rubber-based products useful.

John Bain stretched 850,000 rubber bands into a ball with a diameter of more than five feet. To date, it is the largest rubber band ball in the world.

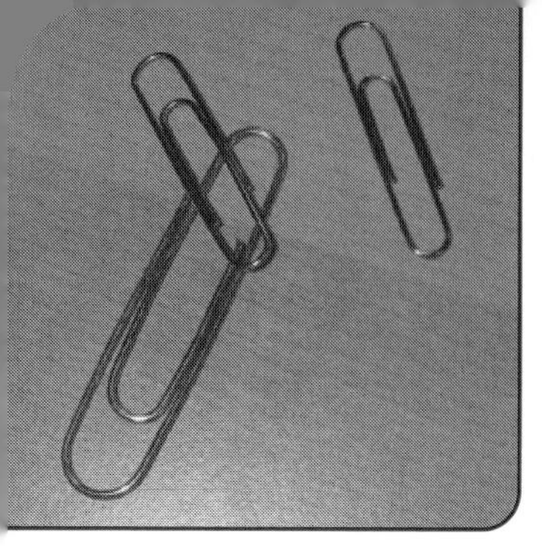

Paper Clip

BEHAVIOR

Holds papers together neatly but releases them with a flick of a thumb.

HABITAT

They reside with many siblings in a box or pocket in the top drawer of office desks.

HOW IT WORKS

A paper clip is a spring. You separate the two rounded ends to insert multiple pages between two coils of the spring. The paper clip rebounds, putting pressure on the papers and holding them in place.

Paper clips are usually made of metal, but some are plastic. They come in a variety of sizes and shapes for different jobs. The most common form is made by coiling the two ends of a four-inch piece of steel wire.

Other forms of paper fasteners include the brass fastener that is inserted into holes punched through all the papers being fastened. The two legs of the fastener are spread apart to keep it in place.

Large metal springs hold larger piles of paper together. Bulldog and binder clips have handles that give your fingers leverage to open the strong spring.

There are many unintended uses for paper clips. Unbent, they can reset the digital clock in your car, open recalcitrant CD drawers in your computer, and reset PDAs. They also provide the ideal amount of weight for paper airplanes.

INTERESTING FACTS

The Gem Manufacturing Company is most often associated with paper clips. In some countries paper clips are called "Gems."

If your travels take you to Norway, look for the world's largest paper clip. It is more than 22 feet long and is a tribute to the Norwegian inventor of paper clips, Johann Vaaler.

Stapler

BEHAVIOR

Inserts small metal clips, or staples, into multiple sheets of paper to hold them together.

HABITAT

Staplers reside on or in most desks in an office. Larger manual staplers and electric staplers are usually in the document preparation area.

HOW IT WORKS

The stapler drives a metal staple through the collection of papers. The force of the hand squeeze or thumping drives the staple to an anvil on the underside of the paper. Curved indentations on the anvil guide the ends of the staple to bend inward or outward, preventing the staple from falling out of the stack of paper.

The stapler has two kinds of springs. Internally, staples are pushed forward by a coil spring. The top of the stapler, the part you whack, is a leaf spring. Pushing on the leaf spring forces a dull blade to shear off the front staple from the glued line of staples inside. Upon release, the leaf spring rebounds, extracting the blade. As the blade retracts, the coil spring pushes the row of staples forward so the next staple is ready to go.

INTERESTING FACTS

The stapler was invented in 1866 by George McGill. More than 150 years later, new models and designs are still being created.

Staple Puller

BEHAVIOR

Removes the staples you thoughtfully placed in a stack of papers.

HABITAT

Except while in use, it is stored in a drawer of the office desk.

HOW IT WORKS

The stapler puller's two curved pieces of metal are inserted under the staple. As you squeeze the stapler puller, the metal wedges itself between the staple top and bottom, and as the puller closes, it bends the prong ends of the staple outward, releasing the papers.

A spring inside reopens the staple puller so it's ready to use again. The puller represents the union of two basic machines: the wedge and spring.

Scissors

BEHAVIOR

Allow users to cut paper without an externally exposed edge. Usually easier to use than a knife and gives a straighter cut.

HABITAT

In almost all office desks and in document preparation areas.

HOW IT WORKS

It's not the sharp edges of scissors that cut the paper; it's the shearing of one blade against the other with the paper caught in between. To get the shearing action, the blades are curved. The curve allows the blades to come together at all points as the scissors close.

Scissors are an example of a simple machine. The hinge or pivot point is a fulcrum for two levers, the blades. Long handles give the user mechanical leverage to apply more force to cutting.

UNIQUE CHARACTERISTICS

Open a pair of scissors and, without poking your eyes, look along one blade. Can you see that it isn't straight?

INTERESTING FACTS

Scissors were one of the earliest office inventions. Ancient Egyptians used them as early as 1500 B.C. Improvements were made over the years, the biggest being the introduction of steel blades in the eighteenth century.

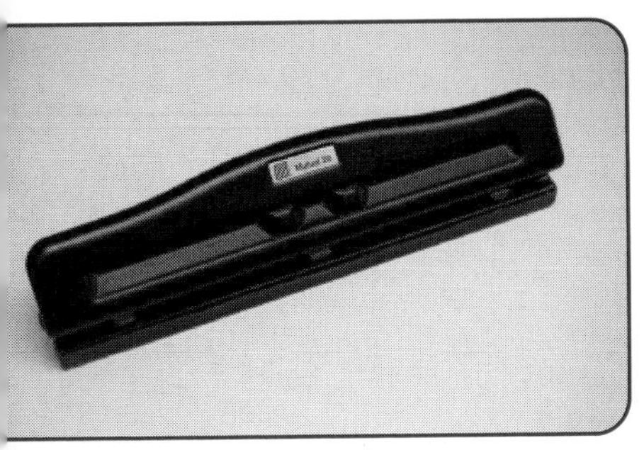

Three-Hole Punch

BEHAVIOR

It produces those annoying small circles of paper that litter the floor around trash cans. It also makes consistently spaced holes in stacks of paper so the paper can be inserted into three-ring binders.

HABITAT

Sits on a table in the document preparation area or in a large desk drawer. You can find it by following the trail of small white dots of paper on the floor.

HOW IT WORKS

It's leverage. You whack the end of a lever, and as it rotates it pushes three sharp punches downward through the paper. Each punch is a steel cylinder with a sharp cutting edge. These cutters are centered 10.8 cm apart for standard three-ring binders. A spring wraps around each cylinder to push it back into position after it has been forced downward, through a stack of paper. This works . . . except, of course, when you've forced more pages into the punch than it can accommodate. Then you have to pry the cutting cylinders back up. You can use pliers or a screwdriver to push them back into position.

Many hole punches have adjustable positions for the punches so you can use binders with other than the standard spacing. Some offices use two-hole punches, and some countries use four-hole punches. But in the United States, the three-hole punch reigns supreme.

Apparently the hole punch was inspired by train conductor's paper punch. Conductors punch holes through tickets to show that they have been used. The office single-hole punch looks much like a conductor's punch.

Adhesive Tape

BEHAVIOR

Used for myriad purposes requiring two items or parts of items to adhere to each other.

HABITAT

Adhesive tape resides in tape dispensers on desks or in one of the top drawers of every desk.

HOW IT WORKS

Adhesive tapes are made by attaching an adhesive emulsion (a mixture of two substances that don't blend together) to a cloth or plastic film.

Surprisingly, there is still great debate about why adhesive tape sticks. A small part of its stickiness is due to electrostatic attraction between the tape and paper. Electrostatic forces are often demonstrated by rubbing a balloon on cloth and having it stick to the ceiling. But recently a new theory has been proposed to account for the stickiness. The tape surface of adhesive materials is not a flat surface; it has deep ridges and valleys. Deep, that is, on a small scale of several ten-thousandths of an inch in height. Pressing down on the tape squishes these ridges and captures molecules of air in the valleys that act as tiny suction cups. Pulling up on the tape is equivalent to breaking the suction of hundreds of thousands of tiny suction cups. So adhesive tape sticks because its stickiness allows it to form vacuum pockets.

INTERESTING FACTS

Adhesive tape works differently than other office adhesives. Glues, usually stored in plastic tubes, harden when exposed to air. Either part of the glue chemical evaporates allowing the other components to harden, or two parts of the glue combine in the presence of oxygen (which is in the air, but not in the tube). Most traditional glues harden on evaporation; superglue components combine when exposed to oxygen.

Post-it Note

BEHAVIOR

It serves as a reminder and page marker. It carries short notes and sticks where you want it to—and unsticks when you need it to.

HABITAT

A packet of Post-it notes is in most office desk drawers.

HOW IT WORKS

Post-it notes are a form of stationery with adhesive on the upper edge of the back of each sheet. The adhesive has relatively weak stickiness, which allows the user to place, remove, and replace a note many times without leaving marks.

The classic Post-it note is a three-inch square of light yellow paper. However, the success of the notes has spurred a wide variety of other sizes, colors, and uses. Some pads of notes are printed with witty comments and illustrations.

INTERESTING FACTS

The fascinating story of Post-it notes is how they were invented. Ubiquitous in today's office, when first created no one had a use for them. Art Fry at 3M searched for uses for a glue invented by fellow 3M researcher Spencer Silver. Fry used the glue to hold book marks in his hymnal. Others didn't see the utility of this product, so 3M gave samples away. Once office workers discovered their own uses for the notes, they started ordering them and Post-it notes became a big hit for 3M. In 1995, 3M was awarded the National Medal of Technology, at least in part due to its invention of the Post-it note.

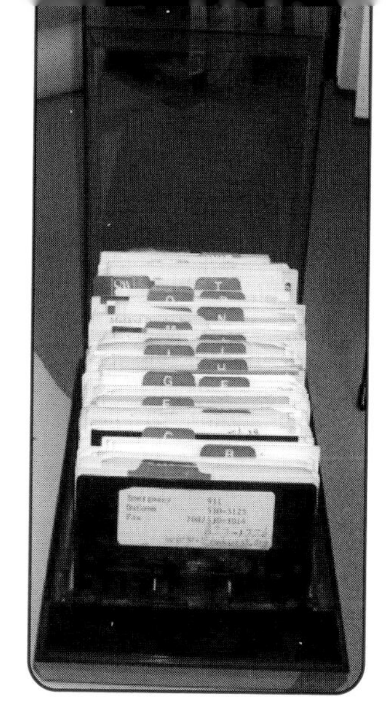

Rolodex

BEHAVIOR

Keeps important contact information immediately accessible on a desk.

HABITAT

Usually on desktops, adjacent to telephones.

HOW IT WORKS

Information is written on and stapled to the specially cut cards that fit onto the rails of a Rolodex. You can remove the cards and return them by pulling them off and pushing them onto the rails.

INTERESTING FACTS

Hildaur Neilson invented the Rolodex in 1938 after having created earlier models. Neilson had several other inventions for the office including the Autodex (a telephone directory that opened to the search letter), a paper hole puncher, and an inkwell that didn't spill. He liked the suffix "-dex," and used it to name several inventions. However, Rolodex was his crowning achievement.

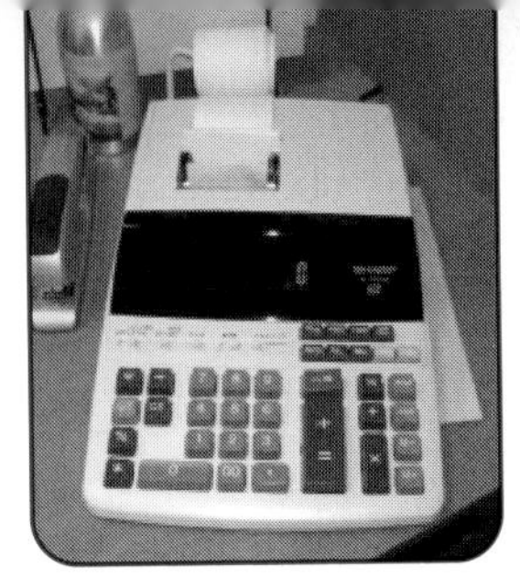

Calculator

BEHAVIOR

It saves the mental tribulations of figuring out who owes what amount for lunch, and performs other digital calculations at tremendous speed and accuracy.

HABITAT

Some models sit patiently on desktops awaiting digital exercise, while others hide in coat pockets or the bottom of briefcases or backpacks.

HOW IT WORKS

Tiny portable calculators get their energy from solar panels. Larger desk models plug into wall outlets, and some models use batteries. You read the results either on a LCD screen or on a thermal printer output.

You enter numbers and operations on the keypad—notice that it has a different layout than the telephone keypad has (see page 101). The numbers are stored in a memory chip. There are three ways to enter information into a calculator: RPN (Reverse Polish Notation, used in old HP calculators), algebraic, and equations. In RPN you enter the numbers and then the operation you want to perform on them. In algebraic modes, you enter the numbers and operators as you would say them: number-operation-number. In equation mode, you enter an equation and store it, adding the variables you want for each calculation.

Digits are converted to binary coded decimal (BCD). In BCD, each digit is represented by four binary bits. The calculations are carried out by integrated circuits.

INTERESTING FACTS

Before calculators came to the market in the early 1970s, people used slide rules. Try finding a slide rule in a store today to see how technological advances change the marketplace. The first personal calculators cost more than $100 and performed only a few basic functions: addition, subtraction, multiplication, division, and square roots. Today, calculators are giveaway premiums and cost a fraction of $100.

Electronic Dictionary/Translator

BEHAVIOR

Provides the correct spelling and meaning of a word, or translates the word into another language, even languages that aren't based on our alphabet.

HABITAT

Resides on the desks of people in international business and in the business of words (writers).

HOW IT WORKS

Type in your best guess as to the spelling of a word, and the dictionary produces the correct spelling and meanings. A typical dictionary has a vocabulary of more than 100,000 words.

These devices are available for a variety of languages and for use as translators. Type in the word in English and out comes the word in Spanish or Italian.

Compact Disk, Read-Only Memory (CD-ROM)

BEHAVIOR
Stores prodigious quantities of information on a 12 cm-diameter, thin plastic disk.

HABITAT
Neatly stored, CDs are stacked and held in place by a plastic spindle that runs through their center opening. Other storage locations include notebook-like containers with plastic pages that have pockets for CDs. Otherwise, CDs lie scattered around the office.

HOW IT WORKS
CD-ROMs are a preferred way of storing software and other digital data. The data can be read only by a computer.

The CD is 120 mm in diameter and 1.2 mm thick. It holds 650 or 700 megabytes of data. The disk is made of polycarbonate plastic covered with a thin layer of aluminum and protected with a coating of lacquer. Information is etched into the aluminum (on the underside of the CD, not the side with the label) using a laser and is read by bouncing a lower-power laser beam off the surface. Data is recorded on a CD in a spiral pattern starting nearest the center opening. The track is about 5 km (3.1 miles) long. The etches made in the CD by the recording laser are 100 nanometers deep—a nanometer is one *billionth* of a meter. To clearly see a pit, you'd have to enlarge a CD so its diameter is wider than a kilometer.

The reading laser inside the CD read/write drive bounces light off the surface of the CD. When bounced off a pit, the reflected light is greatly reduced in strength. Light bounced off the smooth areas reflects strongly. The change from a pit to "land" (no pits) and from land to pits denotes a binary one, while no change in the reflected signal denotes a binary zero.

Typewriter

BEHAVIOR

Transfers finger flicks on a keyboard to letters on a piece of paper. Today typewriters are used mainly for typing postal labels or filling in forms.

HABITAT

Formerly the focal point of the office, these tools of yesteryear are now found mostly in common work areas or in a storage room. Where they once dominated an office with many workers banging away at the keyboards, now it's difficult to find even one typewriter in an office.

HOW IT WORKS

Mechanical typewriters are marvels of engineering. Opening one up or taking one apart gives you a great appreciation for the level of workmanship and engineering. Just picking one up gives you an idea of its mechanical prowess and durability. No wimpy light-weight keyboard here—these machines are *heavy*.

Pushing on a key (that represents a letter, number, or symbol) engages a series of levers, the final lever having the corresponding character at its striking end. The lever arm with the letter moves quickly toward the paper, strikes a roll of inked ribbon, and imparts the image of the letter onto the paper. It was a great innovation that was used in offices for decades.

Adding an electric motor to the typewriter made the keys easier to push and allowed faster typing. Electrics also eliminated key jamming, which occurred often with mechanical typewriters when two keys were pushed simultaneously. The ultimate typewriter innovation was the IBM Selectric, which had the striking letters on a ball, which could be changed to other balls with different fonts.

From electric typewriters, the next step was electronic typewriters. Keystrokes became data bits stored on magnetic cards. One magnetic card could store an entire page of text! As the computer revolution revved up, these evolved into word processors. Personal computers took over from there.

INTERESTING FACTS

The QWERTY layout of the keyboard, introduced in 1874, still dominates keyboards today. No one is sure why, but it seems likely that this layout was selected as the one best able to minimize key jamming in mechanical typewriters. As that is not a problem today with electronic keyboards, more efficient layouts could be (and are) offered, but so many people have learned QWERTY that it continues to be the most popular keyboard layout in English.

Wite-Out or Liquid Paper

BEHAVIOR
Covers up mistakes on documents.

HABITAT
It is usually in the top desk drawer, ready to be called upon to hide smudges and other boo-boos.

HOW IT WORKS
Liquid paper is a paint that dries quickly, covering errors on documents and allowing new information to be written or typed over it. Before the days of computers with spell and grammar checks, and storage for instant retrieval and change, Liquid Paper was a blessing to typists.

The original batches were made from water-based paint. A secretary, Bette Nesmith Graham, wanted a more efficient way to correct minor mistakes. Before memory typewriters came on the market, typists commonly retyped entire pages to correct a single error. Graham experimented with paint and, achieving a color to match the stationery, she began using it at work. Other secretaries asked her for bottles of the stuff (which she called "Mistake Out"), and later she started selling the product.

Graham launched her own company to produce the product and eventually gave up her typing job. The Gillette Company purchased her company in 1976 for some $47 million dollars, a nice piece of change for a single mother raising a child. (Bette's son, Michael Nesmith, was successful in his own right. He was a member of the popular band the Monkees.)

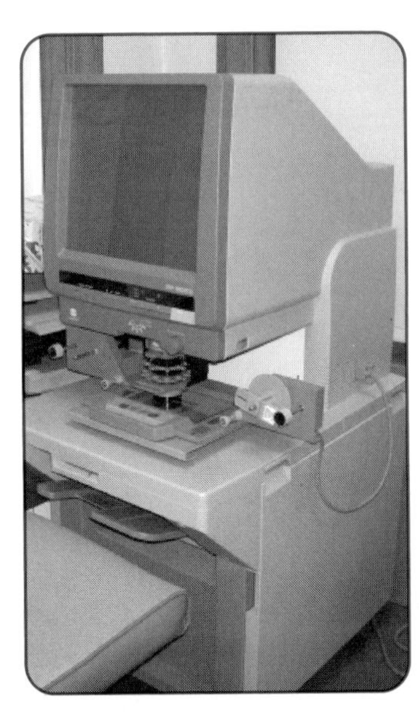

Microfiche Reader

BEHAVIOR

Allows people to read and retrieve information from records saved on microfiche film. The reader shines a light through the film displaying its contents, greatly magnified, onto a screen.

HABITAT

Libraries, both public and private, have microfiche readers. Many sit on a desk adjacent to files holding microfiche films.

HOW IT WORKS

Unlike high-tech data storage devices that store information in digital formats, microfiche systems store information as analog images. That is, a photo of the page is taken and mounted with nearly 100 other page images on a single 4-by-6-inch card. Identification information is printed on one end of the card so each can be found.

A user places the selected card in a moveable tray in the reader, turns on the internal light, and adjusts the focus. To read other pages on the card, the user can move the tray that holds the card above the light.

Microfiche is the most compact analog data storage system used in offices. Huge quantities of data can be stored in a small space. The content of thousands of books fits in a volume about the size of a bread box.

Binding Machine

BEHAVIOR

Punches holes so plastic or metallic material can be inserted to keep all the pages of a document together.

HABITAT

Binding machines are in document preparation areas of offices and commercial copy stores.

HOW IT WORKS

There are several principal types of binding machines. Most require you to punch holes through the collated pages of a document. Some machines hold the pages in place while you insert the plastic or metallic binding piece.

Coil bindings use a continuous coil of wire that you insert into the holes. Comb bindings have a solid plastic back with many teeth that you insert through the rectangular holes punched by the machine. Wire bindings are similar to comb bindings, but the material is made of metal. Thermal binders heat a plastic covering that contains the document to bind the pages together. In a VeloBind system you punch holes through the paper and insert a plastic comb so its rigid fingers poke through the holes. Another piece of plastic is placed on the top page with the fingers protruding through its holes. This is heated to melt the fingers and bind the document.

Perfect binding does not entail punching holes. The machine firmly holds the pages together and glues the pages along the spine (inside edge) of the document.

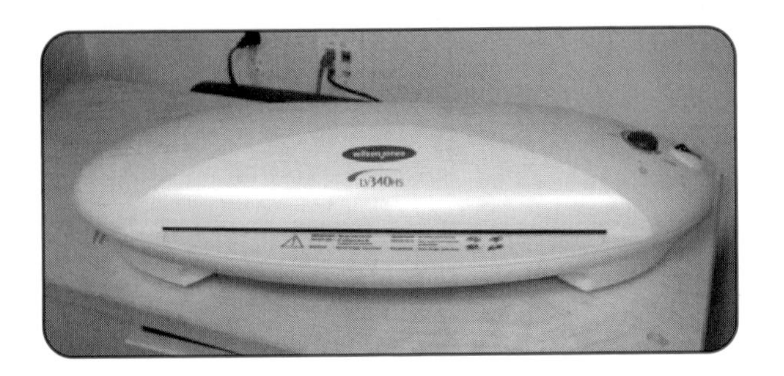

Laminator

BEHAVIOR

Protects individual pages of text or photographs by sealing them in an envelope of clear plastic to keep out moisture and dirt.

HABITAT

You can find laminators in many offices and office product stores. They sit atop a desk in the area where documents are assembled.

HOW IT WORKS

The operator inserts the paper page into a plastic pouch or sleeve. The inside of the plastic (polyester) pouch is coated with an adhesive resin. The pouch is inserted into a cardboard protector and placed in the laminator. Wheels driven by an internal motor pull the cardboard with its contents, the plastic pouch and document. Inside the laminator a heater melts the pouch's resin, which spreads over the document causing the plastic to adhere to the document and to itself outside the boundaries of the document. The pouch passes between rollers that squeeze out air bubbles and help the plastic bond.

The document comes out warm and waterproof. When it cools, the resin bonds the document to the polyester.

Label Maker

BEHAVIOR

Cranks out labels with self-adhesive backing.

HABITAT

Label makers are usually located in document preparation areas or in closets, waiting to be used.

HOW IT WORKS

There is a wide variety of machines for making shipping labels, bar codes, identification tags, labels for computer diskettes, and other uses. Many use thermal printers; that is, they print by heating a special medium instead of by blowing ink at the medium (as inkjet printers do) or electrostatically fixing toner to it (as laser printers do).

Some label makers are stand-alone units with full keyboards, but many connect to computers. Smaller, handheld label makers are called label guns.

There are two methods of thermal printing. In direct thermal printing, the hot print head contacts the special medium and burns the label into it. This requires a special medium and the labels tend not to last as long. In the thermal transfer process, the print head melts the resin on a special ribbon that adheres to a variety of media. This process provides a higher quality print but requires the purchase of ribbons.

Postal Meter

BEHAVIOR

Applies stamps onto envelopes so they can be delivered by the postal service.

HABITAT

Offices often keep their meters in the document preparation area, near paper supplies and the copier.

HOW IT WORKS

Invented by Arthur Pitney in 1912 (who joined Walter Bowes in 1920 to form Pitney Bowes), the postal meter lets you set the amount of postage and mark it on envelopes. The meter keeps track of how much postage you use and, when it reaches its limit, requires a postal recharge at a post office (or over the Internet).

Individuals and companies lease postal meters from one of a handful of companies approved by the U.S. Postal Service. These companies lease the machines to make a profit as they are not allowed to mark up the cost of postage. Older machines require a trip to the post office with check in hand to get the postage recharged. Newer machines can plug into a phone line to get more postage.

Postal meters are electromechanical machines. You set the postage manually, insert the envelope, and press a button. The machine prints a stamp showing the appropriate amount on the envelope and deducts that amount from the amount remaining in its register.

The latest in postal meters has a scale that connects to your computer. You put a letter or package on the scale, and the computer calculates the postage required and prints out a label with the postage. This innovation reduces the postal meter to software and peripherals.

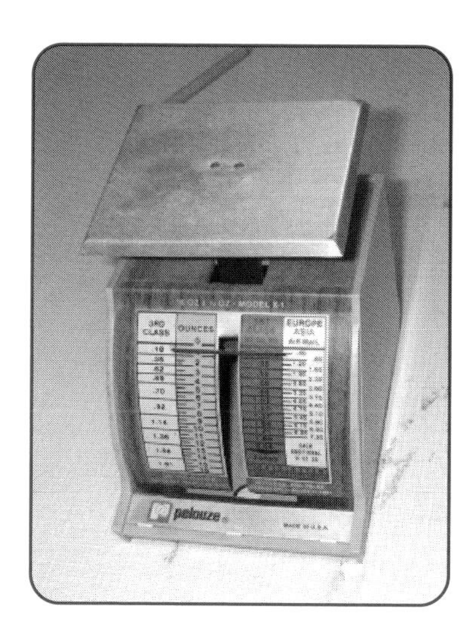

Postal Scale

BEHAVIOR

Measures the weight of letter envelopes and small packages so the correct postage can be applied.

HABITAT

Postal scales reside in document preparation areas, near the postal meter.

HOW IT WORKS

The postal scale traditionally is a spring scale or a balance beam. The spring scale has an internal spring. As a package is placed on the tray, its weight compresses the spring in a linear manner: the greater the weight, the farther it moves. (Hooke's Law, for you physics fans.) The spring is connected to a gauge with calibrations for weight or for postal rates.

The balance beam requires users to adjust weights along a beam to exactly balance the weight of the envelope. Once the weights are adjusted, their position indicates the weight.

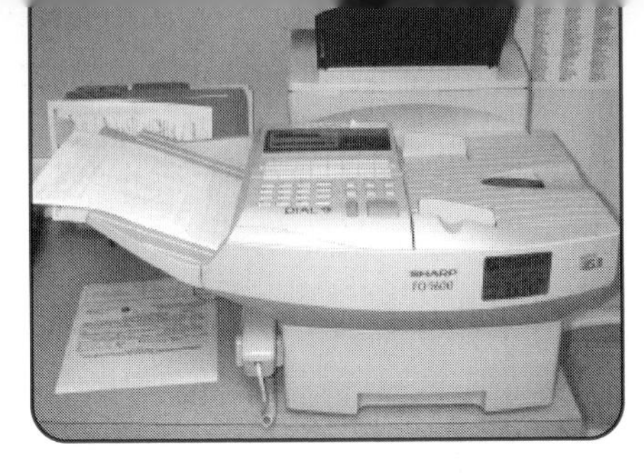

Fax

BEHAVIOR
It converts printed documents into electric signals that can be transmitted through telephone wires to distant machines.

HABITAT
Faxes are usually located near copiers in an area set aside for document preparation.

HOW IT WORKS
Faxes can be stand-alone machines or scanners that rely on computers to send their output. In either case, the critical component of the machine is its scanner.

The fax scanner has more than 1,700 photodiodes to cover the width of a standard piece of paper (8.5 inches). Each one gives a reading of reflected light levels more than 1,000 times down the length of the piece of paper. This sampling results in almost two million bits of data per page. To save time sending and receiving this data, it is compressed using one of several different schemes. Light for the photodiodes to sense comes from a fluorescent bulb. Either the paper is pulled past the bulb and sensors, or the sensors and bulb are pulled over top of the paper.

Receiving an incoming fax, the machine decodes the signal and displays the image as a graphics image, which is sent to the printer.

Stand-alone models connect directly to a phone line that you can see trailing out of the back of the machine along with an electrical power cord. Scanners connect to a computer and not to a phone line. The computer can be connected directly to a phone line or through a DSL or cable modem.

INTERESTING FACTS
The first fax was invented in 1843 by Alexander Bain. It worked with the telegraph system.

Telephone

BEHAVIOR

It interrupts serious thought and brings news, both good and bad.

HABITAT

Virtually all offices have one or more phones, although some may not be connected to traditional land lines. Traditional phones are placed on desktops, very close to where the worker sits.

HOW IT WORKS

Two wires connect the phone to the RJ-11 socket in the wall. One wire is called the *ring* and the other the *tip*. These connect, through the hook switch, to the handset. Picking up the handset closes the hook switch that connects you to the circuit. The microphone in the lower part of the handset converts the sound of your voice into electrical signals that travel to the boss's phone. The receiver converts the electrical signals coming from your boss's phone back into sound.

An electronic device inside the phone keeps you from hearing your own voice. A bell inside the phone alerts you to an incoming call. The touch pad is an array of switches that turn on pairs of sound generators. Each unique combination of sounds represents the number you've pushed. Those sounds are converted to electrical signals that guide your call through the many switches between you and the person you are calling.

INTERESTING FACTS

Alexander Graham Bell's invention of the telephone in 1876 led to the most financially rewarding patent ever issued. Without a working model, Bell filed for a patent hours before rival Elisha Gray, who had a more advanced design, filed his application.

Answering Machine

BEHAVIOR

It intercepts phone calls, identifies you, explains that you are doing something so important that you can't come to the phone, and invites callers to leave a message. You can listen to the recorded messages at your convenience. Most allow you to retrieve messages remotely by pressing a code on the phone pad when you call your own phone number.

HABITAT

It resides next to the telephone or is part of the telephone.

HOW IT WORKS

Like many creative inventions, the answering machine resulted from combining two existing devices: the tape recorder and the telephone.

Older models record both outgoing and incoming messages on small cassettes of magnetic tape. Voices are recorded on the tape with a write head that aligns the magnetic fields of the metal oxides embedded in the tape. A read head interprets the alignment of the magnetic fields and converts them into electric signals that travel to a speaker, where the electric signals convert to acoustic signals.

Newer models use digital recording. The acoustic voice is converted to an analog electric signal. An analog to digital converter transforms the continuously varying signal into a series of binary bits that are recorded in a random access memory, much like the memory in your computer.

INTERESTING FACTS

The earliest telephone answering machine was invented in the 1930s. However, until the 1970s telephone answering machines were not commonly used.

Cordless Phone

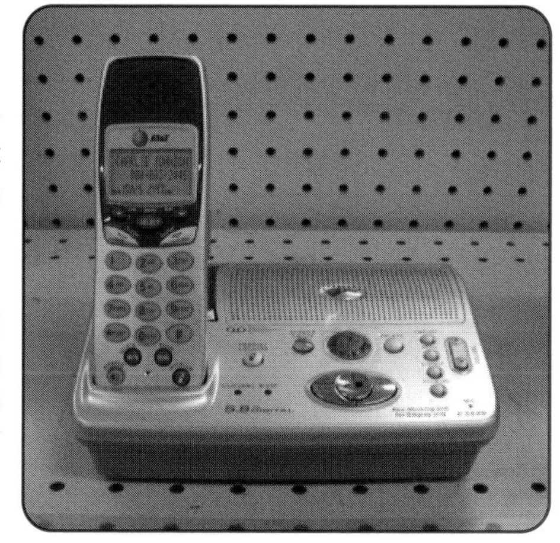

BEHAVIOR
Lets you talk to important clients without being tethered to one location.

HABITAT
These can be found anywhere a telephone is likely to be, but are especially useful for people on the go.

HOW IT WORKS
A cordless phone is a combination of a telephone and two frequency modulated (FM) radio stations. The base receives calls the same way a traditional phone does. It broadcasts the voice as an FM signal to the hand unit. (The signals are broadcast at a frequency that you cannot hear with a traditional FM radio, so that others cannot listen in.)

The hand unit receives the FM signal, converts the radio waves into electric signals, and plays them through a speaker. It also converts the phone user's voice into electric signals and then an FM radio signal that travels back to the base. There the radio signal is converted back into an electric signal and sent out along the telephone wires.

The two FM radios—one for talking and one for listening—operate at slightly different frequencies so the two voices don't interfere with each other. In normal radio operations, such as on a citizen's band (CB) radio, both voices are carried on one frequency, which requires that each user signify when he or she is finished speaking by saying, "Over."

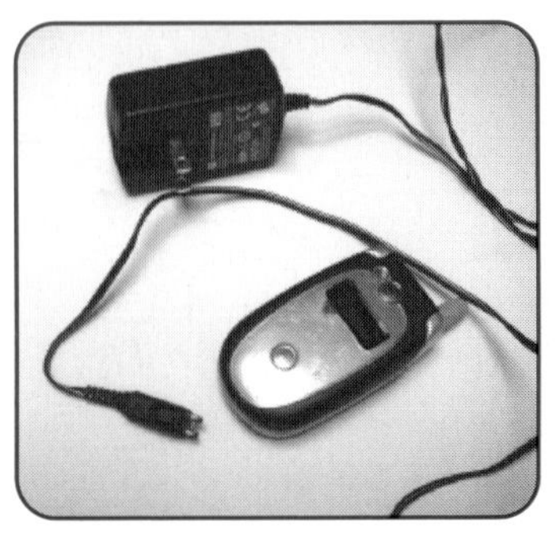

Cell Phone Charger

BEHAVIOR

Replenishes the energy in a cell phone's batteries so the cell phone can be used.

HABITAT

Desktops. Workers can easily plug in their phone to charge while they are sitting at their desk.

HOW IT WORKS

In the United States cell phones seem to have specialized connectors that inhibit you from charging one company's phone using another company's charger. If you find the matching charger, you insert it into the proper receptacle, usually located on the bottom of the phone. Then you connect the other end of the charger wire to a transformer and rectifier, sometimes called a wall wart. These devices convert the 110-volt alternating current from the wall outlet into the voltage required by the batteries, typically 4.2 volts (to charge a 3.6-volt battery). However, the batteries need direct, not alternating current, so the wall wart converts the 4.2 volts of alternating current into direct current.

The batteries recharge quickly. As they recharge, the current from the charger drops, so there is no harm in leaving the batteries plugged into the charger.

Unlike other rechargeable batteries (NiCads, for example), the lithium ion batteries in cell phones do not have a memory effect. In other words, they won't "remember" how much charge was left previously when you charge them. It is better to charge them before they are absolutely dead.

INTERESTING FACTS

It seems like everyone has a cell phone. You probably know of someone who has two. But in Hong Kong, there are many more phones—about 25 percent more—than people.

Walkie-Talkie

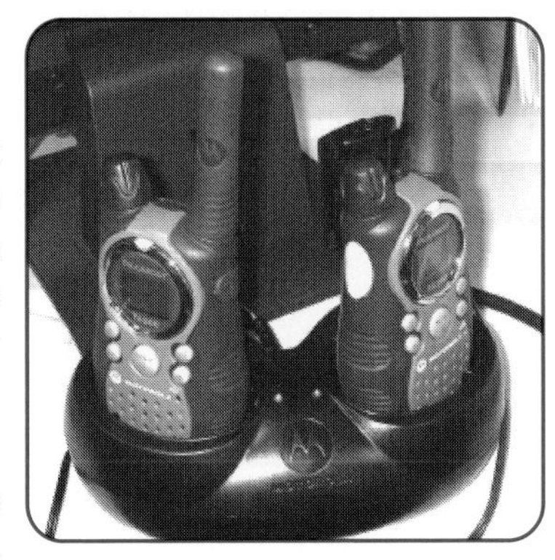

BEHAVIOR

Source of endless amusement and frustration, they allow workers to share information and inside jokes across large offices and manufacturing facilities without using telephones.

HABITAT

Used in businesses where employees move around a lot. The photo above was taken in a children's museum, where staff always run between the museum floor and the office. Shipyards, large assembly plants, distribution centers, warehouses, and large retail outlets all use walkie-talkies.

HOW IT WORKS

Unlike a telephone that offers full-duplex operation, walkie-talkies have half-duplex communications. Both parties can talk and listen, but only one can talk at a time. In half-duplex operation, both parties talk and listen on one frequency.

Each hand unit has both a transmitting and a receiving radio. Walkie-talkies are unlicensed radios permitted to operate over short distances in discrete frequency bands with limited output power.

Most units have rechargeable batteries. To recharge them, the units are set into a charging station overnight.

INTERESTING FACTS

Walkie-talkies are different from Sprint Nextel's walkie-talkie service. It is a modified cell phone service operating on a different frequency than cell phones. A user pushes the "direct connect" button to establish contact. Then the user pushes and holds down the "push to talk" button. He or she releases the button to allow others on the call to talk. Unlike walkie-talkies, users don't have to be in radio range.

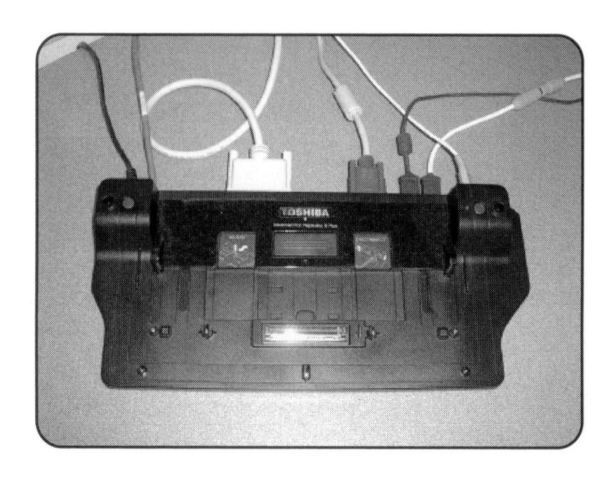

Docking Station

BEHAVIOR

It converts a laptop or notebook computer into a desktop by easily connecting it to a variety of peripheral devices.

HABITAT

On desktops of workers who predominantly use laptop computers.

HOW IT WORKS

Docking stations provide instant connectivity to all the computer accessible devices at the workstation. Someone can use his or her laptop at the desk and have the convenience of a full keyboard, mouse, large monitor screen, printer, scanner, or other hardware—and can take the computer on the road leaving all the other hardware on the desk.

Docking stations and port replicators provide more types of access for the laptop, and they provide that access with only one connection. You plug all the peripheral devices wired into the docking station and then have only one connection to make with the laptop. Docking stations are designed specifically for a laptop model or some handle a variety of models.

Battery Charger

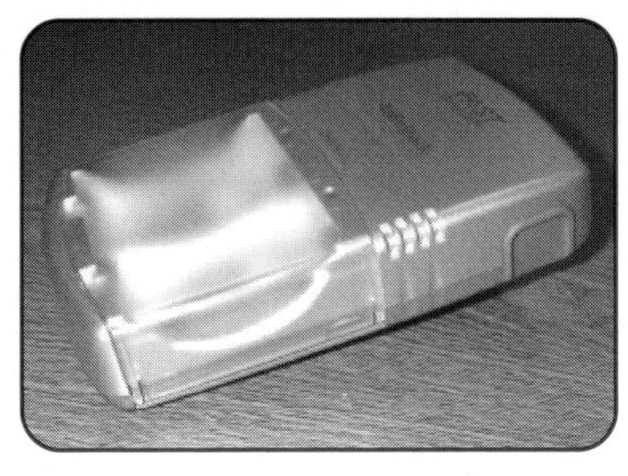

BEHAVIOR
Reenergizes batteries so they can be used again.

HABITAT
On desks, on the floor, or on bookcases.

HOW IT WORKS

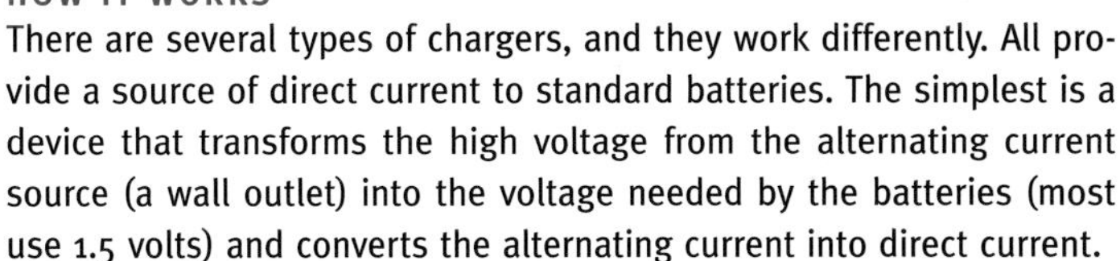

There are several types of chargers, and they work differently. All provide a source of direct current to standard batteries. The simplest is a device that transforms the high voltage from the alternating current source (a wall outlet) into the voltage needed by the batteries (most use 1.5 volts) and converts the alternating current into direct current.

More elaborate chargers include timers that shut off the charging current after five hours (for NiCd, or nickel-cadmium batteries) or eight hours (for NiMH, or nickel metal hydride batteries). Using NiCd batteries in a timed NiMH charger will result in overcharging them, and shortening their life. Using NiMH batteries in a NiCd timer charger will result in undercharging them.

There is also an "intelligent" charger that eliminates these problems. It senses the charge in the battery and reduces the current to the battery as that charge reaches maximum.

"Super-fast" chargers can charge NiMH batteries in 15 minutes. The opposite of super-fast chargers are trickle chargers that continuously supply a low current to charge batteries.

UNIQUE CHARACTERISTICS
When purchasing an appliance, find out what charger the manufacturer recommends for the batteries.

INTERESTING FACTS
NiCd batteries have a "memory" effect: re-charging them before they are drained weakens their ability to hold power. NiMH and Li-Ion (lithium ion) batteries do not share this problem.

IN THE CONFERENCE ROOM

THIS IS A ROOM THAT SITS vacant much of the time. And, although no one likes to see valuable space unused, employees do complain when they have to use it. Yet, the conference room is an essential component of most offices.

The conference room is where executives while away the hours. Though people gripe about attending meetings and how much time they take up, at the end of each meeting they immediately agree to meet again to *really* resolve the problems they were going to resolve at the first meeting.

Conference rooms themselves are often sterile with blank white walls or, for upper echelon meetings, wood paneling. Many are bereft of modern technology, but some have the most expensive devices in the office.

Interactive Whiteboard

BEHAVIOR

Allows groups to write down thoughts and then capture them with a computer. It also projects images stored on a computer. An image can be projected, people can draw and write information about the image onto the whiteboard, and that information can be saved on a computer.

HABITAT

These devices reside in conference rooms of creative technophiles.

HOW IT WORKS

There are several methods of capturing the musings of the group from a whiteboard. Easiest to understand is a pressure sensitive system. Writing on the board forces two sheets of conductive materials to meet at the location of the marker. This contact point is recorded by the computer.

The same approach is taken using an electromagnetic stylus that changes the magnetic field of sensors embedded in the board. The location of the changed field indicates to the computer where the stylus is drawing. Other methods use a rotating laser or infrared light and ultrasonic vibrations to locate the position of writing or drawing. The computer records the position of the marker or stylus and can later display the image.

Overhead Projector

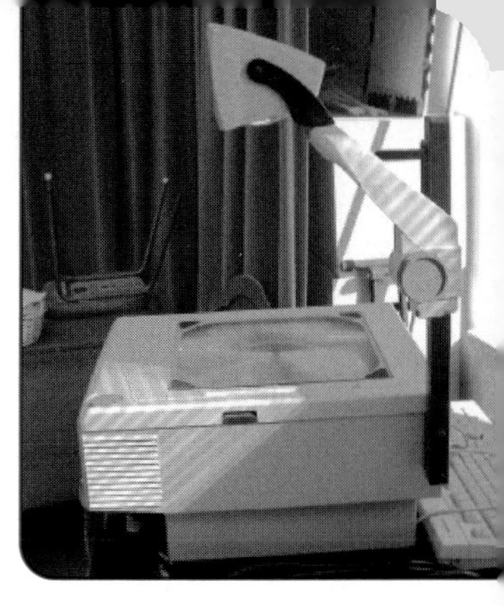

BEHAVIOR

Projects images from transparency copies. Used in lectures and briefings.

HABITAT

Conference rooms and lecture halls. Overhead projectors often sit on rolling carts that are moved into rooms when the machines are needed.

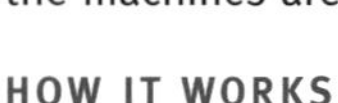

HOW IT WORKS

Pages of information are photocopied onto transparent sheets made of plastic. The copies can be made with standard photocopiers. Alternatively, presenters can write and draw onto transparent sheets for projection. The pages are placed on the bed of the projector.

Below the bed a powerful bulb projects light upward through the glass cover of the bed and through the transparency. The words and images on the transparency block the light. The bulb is so powerful that it generates considerable heat that is dissipated by a noisy fan.

Above the bed, mounted on a vertical arm, are a lens and mirror. The mirror changes the direction of the light toward the screen. The lens collects the light and projects it clearly. The adjustable position of the mirror and lens on the vertical arm allows the user to focus the image on a screen regardless of the distance.

Before overhead projectors, there were opaque projectors. These project the page of a book or an image that is drawn on a piece of paper (not on transparent film). These are used today in offices to project images onto media for artists to trace. Opaque projectors have a powerful bulb above the document to be projected. Light from the bulb reflects off the document into a lens and mirror and onto a screen.

INTERESTING FACTS

Overhead projectors were invented during World War II and were used extensively for training late in the war.

Computer Projector

BEHAVIOR

Projects PowerPoint presentations in vivid colors in front of yawning audiences.

HABITAT

Computer projectors hang from conference room ceilings on platforms, so that they have a clear path to the screen at the front of the room. Alternatively, they are placed on projection stands or carts that are wheeled into different rooms as needed.

HOW IT WORKS

The recent advent of computer projectors has changed the way the world gives presentations and shows images. One of the most popular technologies was developed by Texas Instruments in the mid-1990s. Its digital light processing (DLP) technology works by switching half a million or more microscopic mirrors that control where projected light falls on the screen.

Each mirror controls the light for one pixel or picture element. The mirrors are embedded on an integrated circuit that is called a Digital Micromirror Device (DMD). The mirrors can rotate to reflect light onto the lens for projection or to a "light dump" so it doesn't reach the lens. Color can be projected either using three DMD devices (one for each of the primary colors) or using a spinning color wheel and one or two DMDs.

The other popular technology for digital projectors is liquid crystal display (LCD). Instead of mirrors rotating to cast the light aside or toward the lens, each pixel is controlled by a shutter system that allows light to pass, or blocks it. There are three sets of these systems for the three primary colors. Light is provided by a halogen lamp.

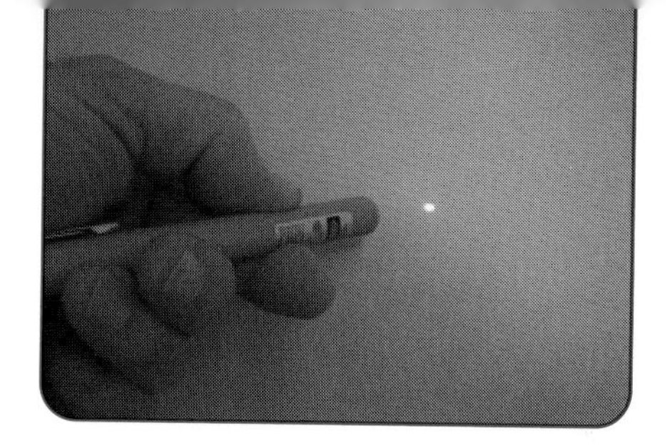

Laser Pointer

BEHAVIOR

Allows speakers to highlight parts of images on projected slides and video.

HABITAT

Spends much time affixed to the speaker's breast pocket. In use, it is handheld.

HOW IT WORKS

Laser pointers are low power (less than 5 mW) diode lasers. They can be of several different colors: red, green, yellow-orange, and blue.

Lasers provide a coherent beam of light, one that diverges very little. So they produce an intense dot of light even over great distance. The intense beam can damage eyes, so laser pointers should never be pointed at people (or other animals).

Diode lasers are similar to light emitting diodes (LED) that emit visible and infrared light. An electric current supplied by a battery causes electrons to move between two different semiconductors. Under lasing conditions, photons are emitted and in turn cause more photons to be emitted. The stimulated emission of light (radiation) is a laser, an acronym for **l**ight **a**mplification by **s**timulated **e**mission of **r**adiation.

INTERESTING FACTS

Robert Hall is credited with inventing the diode laser (1962) used not only in laser pointers, but also in CD players and bar code scanners. He also invented the microwave generator (magnetron) used in most of today's microwave ovens.

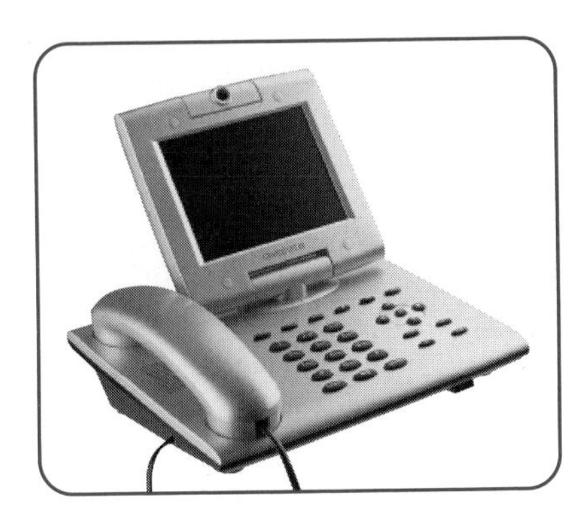

Video Phone

BEHAVIOR

Allows two people who are many miles apart to talk to and see each other at the same time.

HABITAT

Located in conference rooms of large corporations and, as prices fall, in smaller companies and homes.

HOW IT WORKS

Audio and video signals are sent separately over telephone wires. Video signals can come from a camera or other video source. People use video phones for teleconferencing, security monitoring, and monitoring the elderly or children. Signals are sent using a broadband service, such as DSL or cable.

Even a few years ago, the rate of transmission was so low that images jumped and jerked about. This was caused by the low rate of frame capture/transmission. These systems could transmit only a few images each minute, not enough for the human eye to see continuous motion. New designs capture and display images 30 times a second to give steady images.

Individual users can buy video phones to send video and audio signals over the Internet. Companies lease larger systems. With voice-over Internet and a Webcam, inexpensive systems can be set up.

Digital Recorder

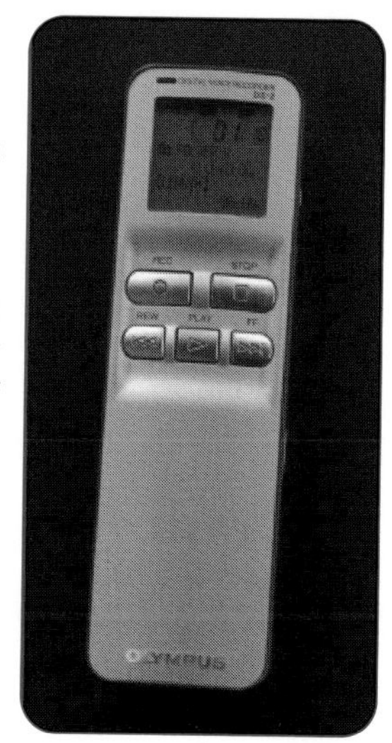

BEHAVIOR

Digital recorders record conference room presentations and conversations.

HABITAT

Modern recorders are small enough to fit inside a pocket or briefcase. In use, they are set on the conference table in front of the presenter.

HOW IT WORKS

Thomas Edison's favorite invention was his audio recorder and record player. This technology was used until wire recorders and tape recorders began to replace it in the 1920s. Wire recorders used hair-thin rolls of wire that moved at two feet per second. To record one hour of sound took about 1.4 miles of wire that compactly rolled onto a three-inch reel.

Tape recorders replaced wire recorders in the late 1940s. Iron oxide on a plastic tape records the sounds. The tape is rolled past a "record head" whose powerful magnets align the iron oxide particles. Playing the tape over a "play head" converts the alignment of the particles back into an electric signal that is converted into sound.

These technologies are analog recording. Digital recording allows storage in a variety of formats (magnetic tape, CD, DVD, flash memory) that won't degrade with age. The digital recorder receives sounds through a microphone and converts them into a digital signal using a special integrated circuit called an analog to digital converter. The converter measures the voltage from the microphone many times each second and records the voltage. During playback, a digital to analog converter reverses the process to generate an electric signal that can be heard through speakers or a headset.

Speakers

BEHAVIOR

Speakers convert electrical impulses into acoustic sounds.

HABITAT

Speakers are usually adjacent to computers and audio equipment. They can be found also in conference rooms. Some are mounted in the ceiling where they play music that no one seems to like.

HOW IT WORKS

The most common speakers are dynamic speakers. They make sounds by vibrating a stiff paper cone, which vibrates air molecules. The vibrations in the air vibrate your eardrum and the tiny bones inside your ear.

The paper cone is attached to a small coil of wire through which the electrical signal, either voice or music, is passed. The changing electrical signal in the wires generates an electromagnetic force in the coil. The force pushes and pulls on a strong fixed or permanent magnet attached at the end of the "spider," the arms that hold the speaker assembly together. Thus, the vibrations in the electrical current cause the coils and the attached paper cone to vibrate at the same rate or frequency as the electric signals. They move air molecules so we can hear the sounds.

A speaker that plays sounds very accurately may have several dynamic speakers inside. Each one responds optimally to a different part of the sound frequency range: woofers (low frequency), mid-range, and tweeters (high frequency). Although some speakers play frequencies above 20,000 Hz, humans can hear in a much smaller range, up to only 15,000 Hz. As we get older this upper limit of hearing decreases.

Other types of speakers are also common. Many of the beeps you hear (when the coffee is brewed or when the microwave oven has heated your taco) are made not by dynamic speakers but by much less expensive piezo speakers. These speakers use special crystals—piezo crystals—that can transform electric energy into mechanical energy and vice versa. The mechanical motion moves air molecules that we hear as sounds.

INTERESTING FACTS

Alexander Graham Bell invented the first speaker as part of his invention of the telephone (1878). Others quickly improved his design and invented other ways to make electrical signals audible.

> Piezo crystals have other common uses. For example, piezo crystals in barbeque grills generate sparks for igniters.

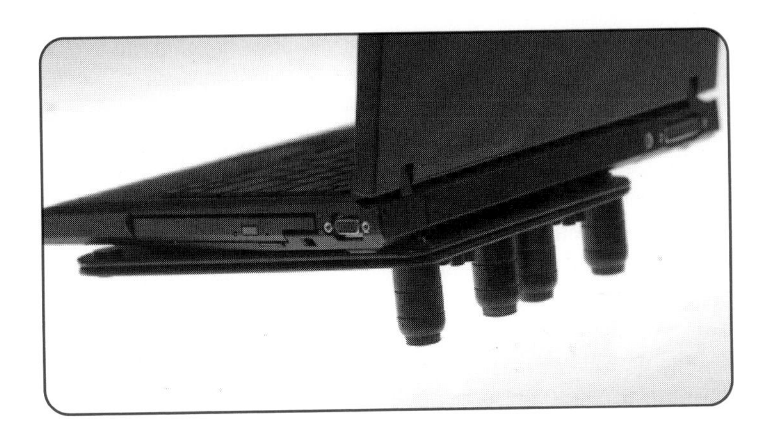

Laptop Cooler

BEHAVIOR
Keeps your laptop cool in the heat of your big presentation.

HABITAT
Sits beneath a laptop while it is performing at podiums or tables.

HOW IT WORKS
There are many variations of laptop coolers, but most share a few common traits.

Coolers help circulate air underneath laptops. To do this, they raise the laptop off the table or desk. Many have electric fans that move warm air away and draw in cooler air. These fans can be supplied electrically from a USB port on the laptop.

Raising the laptop on a few blocks of wood might be as effective, but it certainly won't look high-tech.

Dimmer Switch

BEHAVIOR

Rather than turning a light on or off, a dimmer switch allows intermediate lighting levels. You can set light levels dark enough to see the projected images on a screen and light enough to record notes on a pad of paper.

HABITAT

Dimmer switches can be used anywhere but most often are in rooms where presentation projectors are used.

HOW IT WORKS

Variable resistors are how they *don't* work. In a physics lab you get the voltage you want by using a variable resistor, but this approach wastes power that ends up as heat. You don't want the switch heating up, and you certainly don't want to pay for electricity going to waste. Instead, dimmer switches have an inside circuit that switches on and off 120 times each second. Each time the alternating current changes from positive voltage to negative and from negative back to positive, the dimmer senses zero voltage, and it turns off. It stays off for a length of time dependent on how far you twist the dial.

The knob you turn is a variable resistor. It is connected to a capacitor (an electric device that stores energy). As you turn the knob, you change the resistance, which changes how quickly the capacitor fills up with electric charge. When the capacitor fills, it discharges and that triggers an electronic component—a triac—to fire. Thus, by turning the knob, you change where in the cycle of alternating current the triac closes, allowing current to pass through the switch to light the light.

INTERESTING FACTS

Why don't you see the lights flicker as they turn on and off 120 times a second? Although the power turns on and off, the bulb stays hot; it doesn't cool off enough between cycles for you to notice a flicker.

MENU
RESET/TEST
ENTER
Panasonic KX-P4410

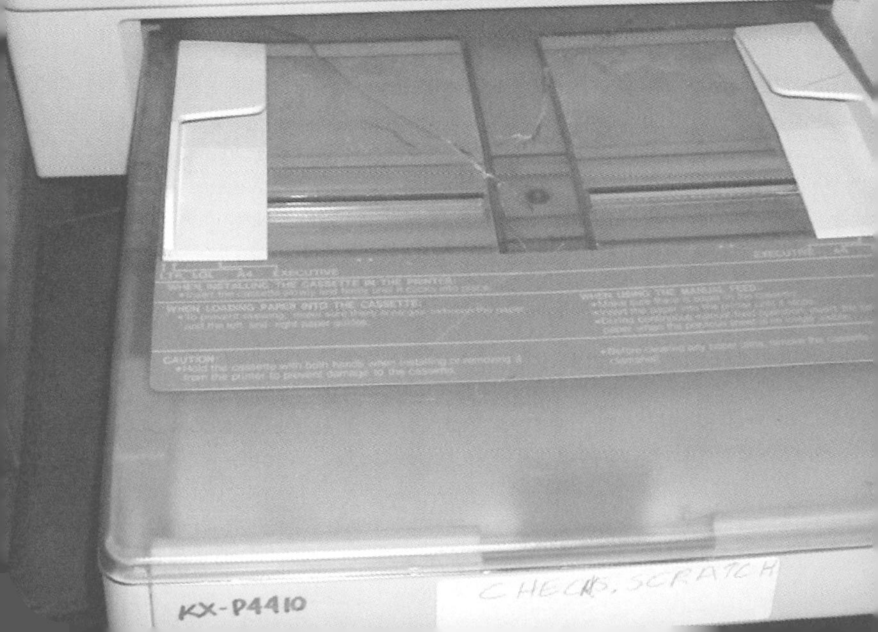
CAUTION:
KX-P4410
CHECKS, SCRATCH

CONNECTED TO A COMPUTER

CAN YOU REMEMBER the early days of the PC revolution? Back in the 1970s and early 1980s, computers connected only to monitors and keyboards. People signed up for classes to write programs for their computers. There really wasn't much you could do with computers, but you had to have one.

Fast forward to that machine sitting on your desk and count all the devices that routinely plug into the computer—or that it plugs into. Computers get their electrical energy not directly from wall outlets but from surge suppressors or power supplies. They connect to printers to make copies of documents, or, if several people share printers, they connect to print servers that connect to printers. Computers write data on CDs, DVDs, memory sticks, and old floppy disks. They connect to the Internet through any of several devices so you can send e-mail or FTP photos of your children or grandchildren. Many computers connect to speakers or headphones so you can listen to your favorite radio station (on the Internet) from anywhere in the world.

Inputs coming into computers come from a mouse, scanners, slide scanners, digital cameras, digital video recorders, and Webcams. With USB and Bluetooth connectivity, it's much easier today to connect many devices to computers.

If there is an ugly side to this potential productivity and usability, it's the mess of wires hidden behind the computer. Maybe this problem will be solved in the next wave of new technology.

The mind-boggling aspect of this connectivity is that it's all done by tiny switches that turn on and off—very quickly. Type in a URL and press "Enter," and in a second or less a computer across the continent has sent you the requested files, possibly by way of a satellite thousands of miles above the earth's surface.

There is amazing technology at your fingertips. Take this guide and scout the neat machines in your office.

Computer

BEHAVIOR

This is the principal tool for many knowledge workers. With human input, it creates documents, calculates budgets, processes drawings and photos, sends and receives messages, and remembers appointments. It also drives many of the other machines that surround it on the desktop and in the rest of the office.

HABITAT

The computer sits center stage on many workers' desks. Some people have two or even three computers. Senior executives place their computers aside so they can clearly see visitors—so they can look 'em in the eye. The most ostensible component of the computer system is the monitor. The computation machine can be located beneath or beside the monitor, inside the desk, or on the floor.

HOW IT WORKS

Personal computer systems are composed of several major components (hardware): a central processing unit (CPU), input/output devices (keyboard, mouse, printer, and monitor), memory (random access memory and hard drives), a clock, a power supply, and wires that connect the components. Computer programs (software) provide directions for processing data.

Computers do very simple operations but do them so quickly that complex functions can be performed much faster than by humans. As hardware speeds have approximately doubled every two years (roughly following Moore's Law), software is more complex and lengthy, allowing more functions and greater details.

Monitor

BEHAVIOR

This is the device everyone stares at when they're working on their computers. Rarely do they stare at the computer itself. The monitor displays the words, numbers, and images in the computer. And it conveys operating information from the computer, such as "file not found" and "virus detected."

HABITAT

Most sit atop the desk. The position of the monitor tells you about the nature of the person's work. If it is located out of the way on an adjoining table or desk, it indicates that the person's primary work is with people, face to face. If it is located squarely front and center, it suggests that computer work dominates.

HOW IT WORKS

Most monitors are either cathode ray tubes (similar to television sets) or liquid crystal displays (found in laptops and some desktop monitors).

A cathode ray tube is an evacuated tube. An electron gun shoots a stream of electrons at the fluorescent screen. The beam of electrons is guided by magnetic or electric fields so it can sweep across the screen and down to form a picture. Traditional television sets work this way.

A liquid crystal display (LCD) starts with a stream of light passing through polarizing filters that sandwich a liquid crystal. Each picture element (pixel) is a cluster of liquid crystal molecules connected to two electrodes. The electrodes for each liquid crystal carry electrical charge to it, causing it to change position and change the level of light it allows past, which feeds the picture signal.

Note that both cathode ray tube models and LCDs are measured diagonally across the screen. However, the advertised size for cathode ray tube models includes the plastic case and the screen; the LCD size does not include the case. Besides clarity of image and depth or number of colors, most important is the size of the viewable screen.

INTERESTING FACTS

Philo Farnsworth invented the cold cathode ray tube used in televisions and monitors. Unlike the majority of inventors in the field of electronics, Farnsworth enjoyed few resources and limited education. Raised on a farm in Idaho, Farnsworth raced and beat the leading companies of the era to develop television. James Fergason, while on the faculty at Kent State University, discovered how to make practical LCDs in 1971.

Keyboard

BEHAVIOR

Allows entry of numbers, letters, and commands into the computer.

HABITAT

Keyboards are often housed either on the office desk or on a tray that slides under the desktop.

HOW IT WORKS

Keyboards are collections of switches arranged in a common format so you can use almost any computer.

Each key is a single switch. Pressing down on one closes its switch, allowing electric current to flow. A microprocessor—in essence a limited function computer—interprets the signals and looks up in a memory bank what the signals represent. Pushing the Caps key and the letter *A* represents something different from pushing the Control key and *A*, and the processor figures out which symbol or action is required.

The cable connecting the keyboard to the computer carries electrical power to the keyboard and the input data from the keyboard to the computer. The most common connection is a universal series bus, or USB.

Wireless keyboards use infrared or radio communications to get information to the computer and require an onboard battery for power. They free the user from being tethered to the computer.

> Look for old keyboards, which are great to take apart. You will be impressed at how simple a keyboard is. And with the keys you take out, you can spell rude messages to glue to your office door.

Mouse and Wireless Mouse

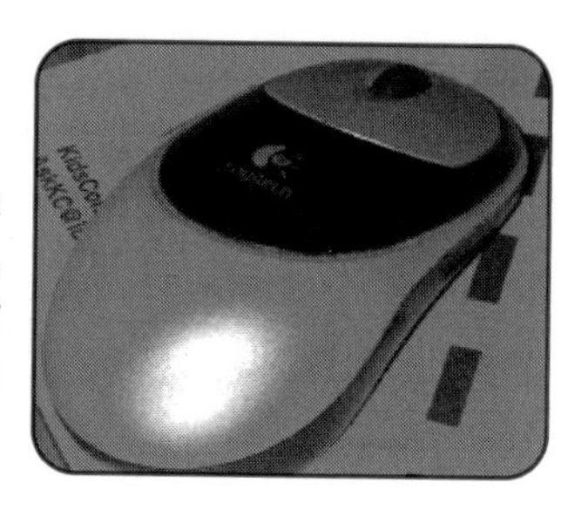

BEHAVIOR

It allows you to move the cursor on a computer monitor and to select words, images, or areas of the display for capture and processing. It also gives you access to some other commands.

HABITAT

It lives on a mouse pad adjacent to the computer keyboard.

HOW IT WORKS

A traditional, mechanical mouse uses a ball that rolls across a mouse pad. As it moves, it rotates two perpendicularly mounted wheels inside. As each wheel spins it turns another wheel with slits. An infrared light emitting diode (LED) shines through the slits and infrared photodiodes count the number of slits that pass. Each passing slit signifies movement in either of the two perpendicular directions.

An optical mouse takes pictures of the underlying mouse pad 1,500 times a second. Changes in the image indicate motion. An LED shines down on the mouse pad and photodiodes record levels of reflected light that indicate the position. These don't work well on reflective surfaces. Increasingly, mouses are wireless. They transmit position information either by radio (Bluetooth technology) or by infrared light.

UNIQUE CHARACTERISTICS

Grab an old mouse on its way to the landfill and take it apart. It's an easy dissection; there are only a few screws holding it together. Look inside and use the description above to understand how it works.

INTERESTING FACTS

You might call it a "mouse," but the original patent called it "an x-y position indicator for a display system." Douglas Engelbart of Stanford Research Institute invented the mouse in 1963. Several names were suggested for the prototype mouse. One was bug. But clearly the pointing device looked more like a mouse, with its extra long tail, than a bug.

Surge Protector

BEHAVIOR

Distributes electric power to several appliances and protects them (and you) from sudden surges in electricity. This low-cost device protects thousands of dollars of important equipment.

HABITAT

One end of the surge protector is connected to computers and peripheral equipment, and the other end is plugged into a wall outlet. Surge protectors are tucked out of sight (or not so out of sight) on the floor beneath desks and workstations.

HOW IT WORKS

Inside the surge protector are several components that divert excess electric power (voltages above 120V) to a ground wire. They also have a fuse in case the diverted circuit doesn't work properly. Some protectors also have components to average out small fluctuations in the voltage.

One way to divert excess current uses semiconductors that have variable resistance. Their resistance is quite high at low voltages and decreases as voltages increase. In normal operation they present such a high resistance that electric power doesn't pass through them, but bypasses them to the computer. But, when a surge occurs, the higher voltages pass through the variable resistors to the ground. The device that performs this voltage switching resistance is a metal oxide varistor, or MOV.

Minor fluctuations in the voltage are suppressed with a "choke," or electromagnet. Current passing through the electromagnet sets up a magnetic field that resists changes. As the current changes, the magnetic field made by the choke induces an equal and opposite change in the current, thus balancing or at least reducing the fluctuations.

INTERESTING FACTS

By definition, a surge lasts at least three nanoseconds and a spike lasts only two nanoseconds or less.

Uninterruptible Power Supply (UPS)

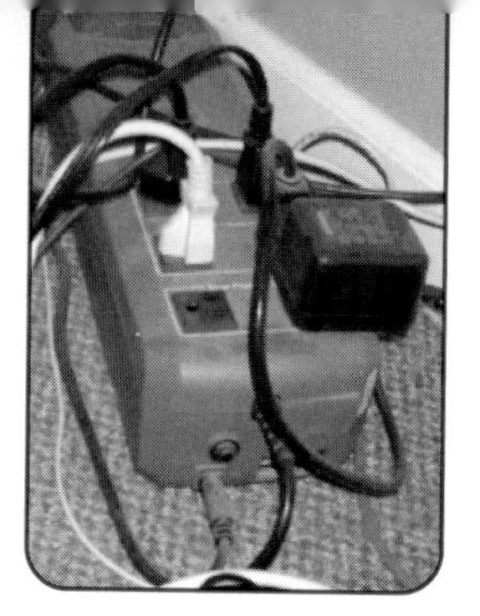

BEHAVIOR

Provides a steady supply of power to computers and other equipment even while spikes, surges, and short power outages occur in the power supply from the wall outlet. This is the optimum in protection from fluctuations in the power supply.

HABITAT

Plugged into a wall outlet and attached to computers and other sensitive electronic equipment.

HOW IT WORKS

A UPS stores electrical energy in a lead acid battery, like your car battery but smaller. Alternating current from the wall outlet constantly charges the battery through a power supply (a device that transforms the voltage to the voltage the battery needs and converts the alternating current into direct current). The battery provides direct current to an inverter that makes alternating current from direct current, and this alternating current supplies the computer with the power it needs.

Less expensive UPS systems operate on a standby mode. They provide alternating current from the wall outlet to the computer unless there is a power outage. Within a few milliseconds of an outage, power begins to flow from the battery to the computer.

In more expensive systems, known as continuous UPS, the computer always gets its power from the battery. When a power outage occurs, the computer experiences no interruption since it was already drawing power from the battery.

INTERESTING FACTS

There is an additional benefit in using UPS: it isolates the computer from the many small voltage fluctuations that occur throughout the workday. Some experts suggest that these fluctuations damage hard drives, so using a UPS prolongs the life of hard drives.

Inkjet Printer

BEHAVIOR

It spits tiny drops of ink onto paper so you can share great ideas, photographs, and business plans with the world.

HABITAT

In many offices, adjacent to the computer. In offices where several people share a printer, the printer is usually located in a common area or document preparation area.

HOW IT WORKS

From one (black), and sometimes three additional (color), reservoirs inside the printer, ink is thrown onto paper. Each dot of ink is smaller than the diameter of a human hair. With amazing precision the inkjet printer can fling these dots to render images almost as precisely as traditional photographs.

Inkjet printers use two methods to shoot ink. One is called the thermal bubble or bubble jet technique. Tiny resistors (300 to 600 of them in one print head) heat up when electric current flows through them. As they get warmer, they heat adjacent ink. The ink vaporizes, expands, and forms a bubble that bursts through a nozzle aimed at the paper.

The piezoelectric method uses piezoelectric crystals that vibrate when voltage is applied to them. As they vibrate in one direction, they draw in ink from the reservoir. Vibrating in the other direction, they shoot the ink out a nozzle.

A stepper motor pulls paper from a tray into the printer. A second stepper motor precisely moves the print head back and forth as the paper moves beneath it. Stepper motors don't move like common motors that spin when current is applied. Steppers take one step, typically two to three degrees of rotation in size, at a time, and thus are ideal for making fine adjustments in position.

Laser Printer

BEHAVIOR

It prints your next novel in clarity not possible even a few years ago.

HABITAT

Laser printers are adjacent to computers in the office or in the document preparation area. They are typically not present if an inkjet printer is there.

HOW IT WORKS

Laser printers evolved from photocopiers and operate much the same way. The data to be printed comes from the computer to the printer where it is stored until the page is ready for printing. A laser draws the images of letters or graphics on a rotating drum, which creates tiny positively charged areas on the drum.

Instead of ink, laser printers use toner. Toner is a messy dry powder consisting of plastic beads with bits of color attached. The toner has a negative charge so it will adhere to the areas of the drum that are charged positively.

A wire, called the corona wire, gives a positive charge to the drum. (Have you noticed how paper and especially overhead transparencies have such a strong electrostatic charge when they come out of a laser printer?) The drum, with toner, rotates to contact the sheet of paper. The toner adheres to the paper, which has a stronger positive charge than the drum.

The paper, with toner adhering to it, then passes through two rollers that press the toner and heat it. The paper comes out with the toner fixed as a printed page.

INTERESTING FACTS

Although laser printers are more expensive than inkjet printers, toner is much less expensive than inkjet replacement cartridges, so laser printers can be more economical to operate.

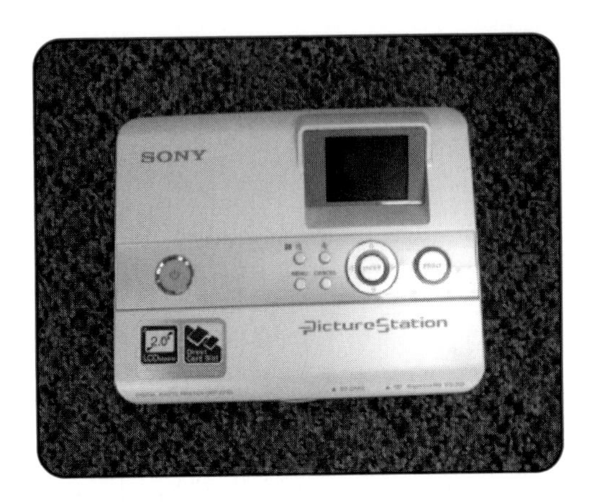

Photo Printer

BEHAVIOR
Prints high-quality photo images from any of several digital media storage devices.

HABITAT
On desks or tables in document preparation areas. Photo printers reside in offices where digital cameras are used extensively.

HOW IT WORKS
Most photo printers are inkjet printers that have more jets spraying smaller quantities of ink than standard printers. The finest scale spits out one trillionth of a liter of ink. As the size of the jets and the quantity of ink decrease, the image quality generally increases.

Connecting the printer to a computer lets you print images saved on the hard drive. However, images can also be transferred directly to the printer from a digital camera by removing the flash memory card and inserting it into the printer. Some photo printers have a cable that connects to the camera.

Modem

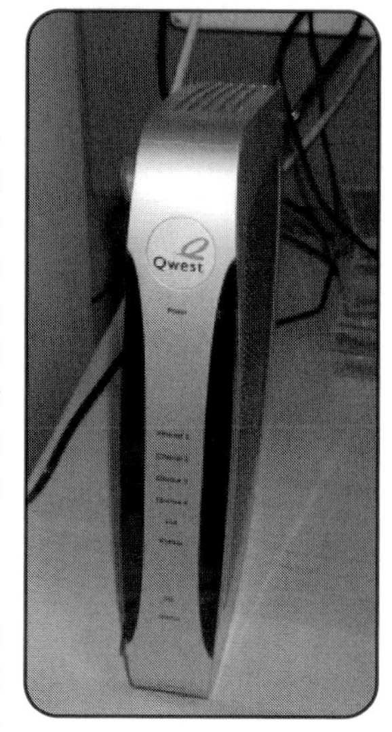

BEHAVIOR

Modems allow computers to pass information to telephone lines for transmission to other computers. They transform the digital signals processed by computers into an analog form carried by telephone wires. They do this by modulating and demodulating the signals that travel by telephone wire and cable. Modems also connect several computers in an office to one high-speed Internet connection.

HABITAT

Modems are usually adjacent to computers on desktops. They plug into computers and into phone lines or cable lines, plus electrical power outlets. Larger modems can be mounted in equipment racks in closets near other communications equipment and connections.

HOW IT WORKS

This device modulates and demodulates a carrier signal with the data to be sent. The resulting signal carries information that can be converted by another modem at the destination into a format usable by computers.

Dial-up modems allow computers to transfer information over telephone wires at voice frequencies, up to 3,400 cycles per second, at a relatively slow speed. The asymmetric digital subscriber line (ADSL) is faster. Asymmetric refers to the differences in the rate of speed in uploading and downloading data; ADSL operates much faster downloading to an office computer than uploading. This asymmetry results from a much broader bandwidth being allocated for downloading than uploading. You can use the same telephone wires simultaneously for talking and surfing the Web because voice communications occur in a different part of the frequency spectrum (under 4,000 cycles per second) than is used for ADSL operations. To use ADSL you must be within about three miles of a telephone central office.

Cable modems encode signals at radio frequencies to send and receive over the television cable system. The same space or bandwidth (6 MHz) is allocated on a cable system for Internet downloads as for the comedy channel, ESPN, or any other TV channel. For uploading information, 2 MHz bandwidth is allocated. Thus, downloading is much faster than uploading.

Satellite modems allow computer communications using satellites. Some systems let you download from satellites but require you to use a dial-up line to upload.

INTERESTING FACTS

The word *modem* is a combination of *modulation* and *demodulation*.

Router

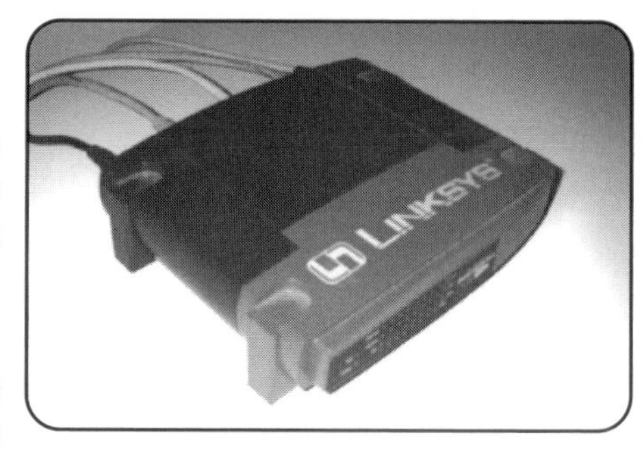

BEHAVIOR

A router connects two or more networks so data packets can reach their destination.

HABITAT

Small routers are on desktops, connected to PCs. Larger routers are housed in separate rooms or tech closets, possibly mounted in equipment racks.

HOW IT WORKS

Routers are specialized computers that steer data packets between networks, groups of computers that share information. Routers sit between networks, examining each e-mail or other data to determine which networks need it. Part of the job of a router is to keep network traffic clear of data packets that don't need to go onto that network.

Unlike a switch on a traditional communications network (such as a telephone) that routes the call one time, routers perform their functions many times for each e-mail or Web page viewed. Internet data is transferred in packets of about 1,500 bytes each, and each one has to be routed from sender to receiver. Each may follow a different path, and the entire communication has to be reassembled at the receiving end in the correct order. Think of how many packets are required to send that photo of your pet or the annual budget spreadsheet!

Routers also provide security from computers outside the network.

INTERESTING FACTS

William Yeager created the first router system (multiple protocol router software) while at Stanford University in 1980.

Hub or Ethernet Hub

BEHAVIOR

Allows one or more computers in an office to share files and to connect to the Internet.

HABITAT

The hub itself sits atop or adjacent to the computer and near the DSL or cable router.

HOW IT WORKS

Special cables, called CAT 5 cables, connect the router to the hub and connect each of the computers to be networked to the hub. These cables are capable of handling much faster rates of data transmission than telephone cables (CAT 1) can. CAT 5 cable has an unshielded, twisted pair of copper wires with RJ-45 connectors at each end. These connectors look like, but are larger than, the standard telephone jack, RJ-11.

Hubs permit connections between computers and with the Internet. They receive signals through one of the ports and repeat that signal to all the other computers that are connected. Hubs are multiport repeaters. Hubs can only handle one stream of data at a time, so if two streams of data arrive at the same time, the sending computers must resolve the conflict through a programmed protocol. Hubs also allow several computers to share peripheral devices, such as printers.

UNIQUE CHARACTERISTICS

Ethernet hubs transmit data at either 10 Mbps or 100 Mbps, the latter called “fast Ethernet.” Mbps is a unit of transmission speed, a million bits of data per second. Most hubs come with four, five, or eight ports to which computers and peripherals can connect.

Wireless Access Point (WAP) Router

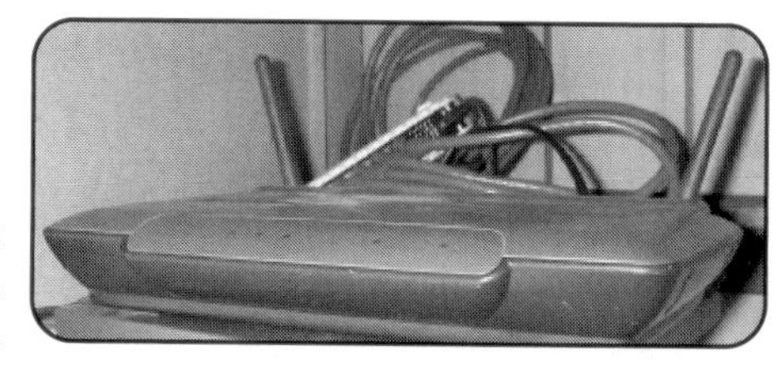

BEHAVIOR

It relays data between the wired network connection and the wireless user. It provides computer access to the Local Area Network (LAN) within an office.

HABITAT

The wireless router is located adjacent to a desktop computer. Laptops or other desktops (with Wi-Fi cards) can be used anywhere in the office or even outside. Public hotspots have WAP routers.

HOW IT WORKS

Easier than running wires through an office, setting up a wireless router makes every room connectable to the Internet and to peripheral devices. Wireless networking, also called 802.11 networking to differentiate it from other wireless systems, requires an access point that is connected to the Internet by wire. It transmits radio frequency signals to wireless-enabled computers up to 100 meters away. The 802.11 refers to the Institute of Electrical and Electronics Engineers (IEEE) standards specified for this technology that uses 2.4 GHz and 5 GHz (gigahertz, or billion cycles per second) frequencies to transmit and receive data.

The wireless router performs three separate functions. It is a hub and connects to the DSL or cable modem. It contains a firewall to provide a primary level of security to keep intruders out of your computers, and it may provide security to prevent unauthorized people from using the system. And it provides wireless connections to up to 30 computers.

To extend the range of a WAP, you can install an extender antenna. Alternatively, you can set up other WAPs in an office accessing the same or different networks. The extenders have amplifiers for both the "send" and the "receive" signals and a taller antenna for better reach.

Peripheral Switch or Server

BEHAVIOR

Allows one or more computers to connect to one or more peripheral devices, like printers, scanners, and more.

HABITAT

In offices where several people need to share the same peripheral devices.

HOW IT WORKS

Older switches were mechanical devices that allowed users to connect several computers to one printer using parallel connections. Newer switches connect with USB devices and operate under a software system that controls the routing of signals. Some systems allow all of the connected computers to print to any of the connected printers without routing instructions other than the printer selection on the computer doing the printing.

UNIQUE CHARACTERISTICS

You can find a variety of capabilities in switches. Start your search by counting the number of computers and number of peripheral devices you might want to interconnect. Then search for a switch that bears the numbers *n* by *m*, where *n* is the number of computers and *m* is the number of peripherals. Most common switches connect two or four computers to one, two, or four devices.

Scanner

BEHAVIOR

It converts printed images and documents into digital signals that can be processed by computers and other machines. Once scanned, information and images can be reproduced, transmitted, changed, or stored.

HABITAT

Scanners usually sit atop a desk or table. Flatbed scanners are placed near the computers to which they are connected. Handheld scanners offer more mobility but still are connected to computers.

HOW IT WORKS

An image or document is placed facedown on a glass plate that lies underneath the lift-up lid. A strong light illuminates the image from below. Three charged-coupled devices (CCDs) scan the page mechanically, sensing the light levels in three different color bands throughout the page.

The charge-coupled device is also used in video and digital cameras. The device consists of an integrated circuit of light-sensitive capacitors. Light falling on the circuit causes charges to be stored in the capacitors, and these charges are measured and recorded. Color images are generated by bringing together the information from the three CCDs, each of which records light in a different color band.

INTERESTING FACTS

In addition to flatbed scanners there are drum scanners. Drum scanners provide more precise images and are more expensive.

CCDs were invented at Bell Labs by Willard Boyle and George Smith in 1969.

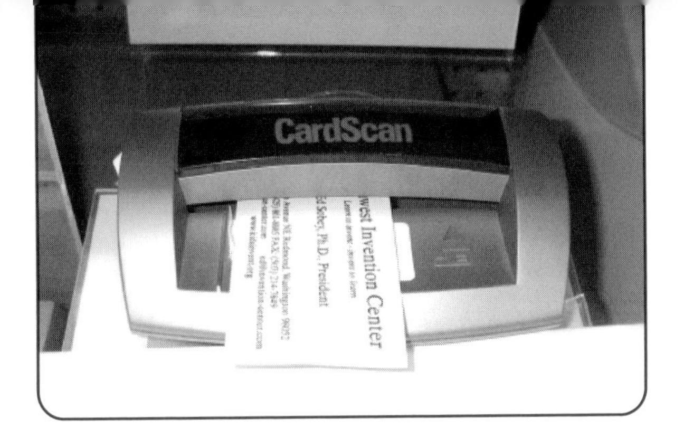

Business Card Scanner

BEHAVIOR

Collects contact information from printed business cards and stores that information electronically.

HABITAT

On desks, connected to computers.

HOW IT WORKS

Insert a business card to turn on an internal light. As the card moves in and out of the scanner, photo detectors pick up the contact information. Optical character recognition (OCR) software converts the images into letters and numbers. Other software sorts and files the information into searchable fields. The scanning process is very quick, allowing users to record information from dozens of cards in a minute.

Once the information is in digital format, users can save it on their PDAs and computers, and can access it with their e-mail programs and even telephones.

INTERESTING FACTS

The U.S. Postal Service has used OCR technology since 1965 to sort mail. The technology evolved from a need to process information gathered for intelligence agencies. Of the several inventors who contributed to the development of OCR, Jacob Rabinow is the best known.

VOIP Headset

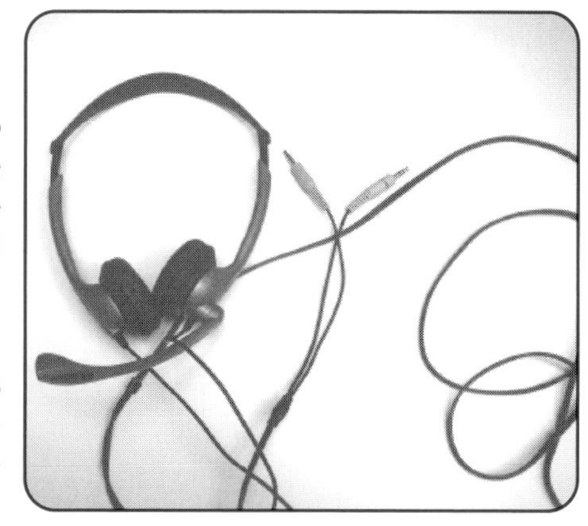

BEHAVIOR

Allows users to talk to distant people over the Internet using a hands-free microphone and headphones.

HABITAT

The VOIP headset is connected to computers, usually in a tangle of wires on a desktop.

HOW IT WORKS

The headphones can encircle the ears, rest on top of the ears, or go into the ears. Each style transforms the electronic signals of voice into sounds. Most systems have dynamic speakers, which use static magnetic fields and electromagnetic fields induced by variations in the electric current (the sounds coming in).

The microphone performs the opposite service, converting audio signals into electric signals. There are several different systems for doing this.

Voice over Internet protocol, or VOIP, allows low-cost calling between two more computers connected to the Internet. The irony is that the calls could travel over the same phone lines that the traditional, more expensive voice calls do. However, with a traditional call, a circuit is dedicated to the users and is thus not available for anyone else to use. With VOIP, the conversation is broken into many very short segments that are sent in packets. At the receiving end, the packets (which may travel by very different routes between sender and receiver) are assembled in the proper order. With high-quality service, users are not aware that they are using an Internet service.

Headphones

BEHAVIOR

They convert electrical signals into sound that the user can listen to without having others nearby hear. Users can listen to music, ball games, noise-canceling sounds, and more.

HABITAT

The headphone jack is plugged into the computer or into a CD player, cell phone, iPod, or other source of music.

HOW IT WORKS

Headphones are a type of transducer, a device that converts energy between two different forms. Headphones convert electrical signals into acoustic signals.

Several mechanisms do this conversion. Most common are dynamic drivers that use an electromagnet moving in the field of a permanent magnet. Changing electrical signals flow through a coil. As the signal changes, it creates a magnetic field that is attracted to or repulsed by the surrounding permanent magnet. The coil is attached to a diaphragm, which moves with it and generates sound waves by moving air molecules.

INTERESTING FACTS

A new use for headphones is to cancel out surrounding sounds that the user finds annoying. They use active noise control rather than trying to insulate the user's ears from sounds. In active noise control, a microphone picks up sounds and a microchip generates an equivalent sound to cancel the first. The two sound waves cancel themselves in the headphones, providing a quiet environment.

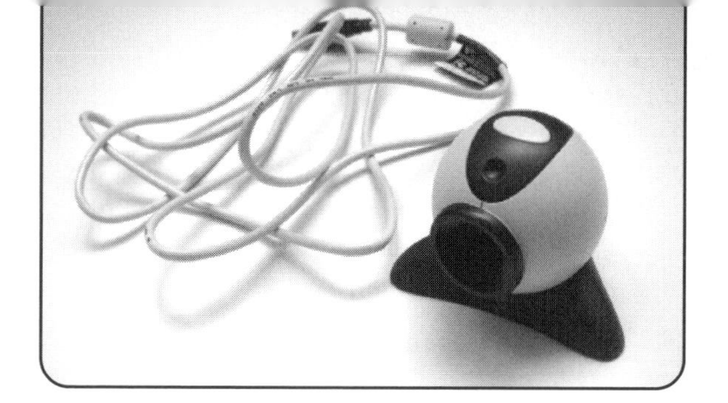

Webcam

BEHAVIOR

Captures video images digitally and sends them to a nearby computer.

HABITAT

Often mounted on top of computer monitors so the desk occupant can transmit images of him- or herself to another person. Others are used at home and aimed at babies, aquariums, pets, and other cherished things.

HOW IT WORKS

These tiny video cameras capture images in low resolution digital format. The camera has a lens to capture the optical image and focus it onto the integrated circuit that detects light and converts it into electric signals. The light generates an electrical signal wherever it hits the surface, and this signal is recorded.

Most Webcams collect the array of electrical signals 15 to 30 times each second and send each one individually to a computer. The more frequently the Webcam collects and sends images, the less jerky the resulting image will be. The computer can compress each image and send it over the Internet.

INTERESTING FACTS

Check out the Internet for Webcams that are transmitting images. Some groups install cameras in bird nests or zoo cages. Museums allow you to see real-time images of their exhibits through Webcams.

USB Light

BEHAVIOR

Provides light for work without having to plug a light fixture into a wall outlet.

HABITAT

Perched on the top of a computer, like Snoopy atop his doghouse, a universal serial bus (USB) light provides a low level of light suitable for lighting a small area.

HOW IT WORKS

Computers provide 5 volts of power (up to 100 microamps, or ma, of current) through the USB ports to power USB devices. The USB light uses this readily available source of energy. Other small appliances, including small vacuum cleaners, fans, and battery chargers, also can be powered through a USB port.

UNIQUE CHARACTERISTICS

USB lights do not have the integrated circuits other USB appliances have that tell the host computer what the current demands will be. USB lights, or fans, can draw too much current and cause problems. Read the package to ensure that the light has a current draw of less than 100 ma.

Digital Pen

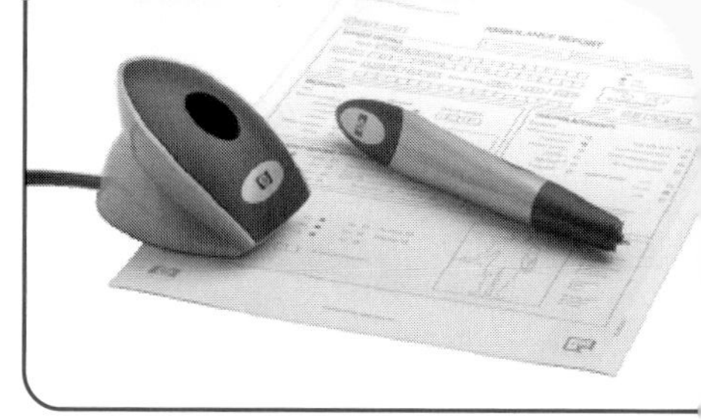

BEHAVIOR

Captures scribbles and written text, saves them in a memory chip, and later transmits them to a computer.

HABITAT

On desktops, near a computer, or safely tucked away in a desk drawer. Digital pens are more common in offices involved with health care, manufacturing, and financial services.

HOW IT WORKS

A digital pen looks like a wide ink pen and leaves a track of ink on the page like an ink pen. The difference is that it also holds a tiny camera and circuitry to capture the motion of the pen.

The camera captures images 100 times each second and sends them to an onboard computer. The change in location of the pen is stored and transmitted wirelessly with Bluetooth technology.

The paper used with digital pens has tiny printed dots or a grid that provide landmarks for the pen's camera. As the pen moves, the camera detects the landmarks that indicate how far the pen has moved.

The pen carries a rechargeable battery. Its memory can store up to 1 MB of data, which is the equivalent of about 40 typed pages. After data is collected it is sent to a computer for processing or recording. Each pen has its own password, so if lost or stolen it is unusable.

The immediate use of digital pens is filling out forms. The forms are printed with the tiny grid. Users write their information on the forms and it is stored in the pen until uploaded to a computer.

INTERESTING FACTS

Using more traditional methods, converting a page of written information to a digital format costs about $1 per page. By using a digital pen, the cost drops to about a quarter of that.

COPPER WIRE &

8 CABLES AND WIRES

DIGITAL AGE ANXIETY RISES when you have to take your computer away from your desk for maintenance. That tangle of cables that was hidden from sight now rears its ugly head. Do you carefully label every wire and where it connected? Or do you think you can figure it out when you bring the computer back?

Cable technology has progressed in step with the other elements of the computer family to generate a bewildering array of wires, each with its own name and function. Specialized cables exist for video download, printer and monitor access, and digital cameras. In some cables the wires run parallel to each other, while in others they are a twisted pair. Still others have one cable inside the other.

The good news is that cables are getting smarter. Older cables are wires wrapped in insulation, hopefully with idiot-proof jacks on each end. Newer computer cables are wires wrapped in insulation with integrated circuits in the end plugs that make connecting machines much easier.

There's no reason to shy away from cables. Step up and master cable terminology to blow away the IT folks.

Electric Cable

BEHAVIOR

Conveys electrical power from wall outlets to a variety of devices.

HABITAT

Since most machinery in an office requires electricity, electric cables are found everywhere.

HOW IT WORKS

Inside most electric cables and extension cords are two or three metal wires that conduct electricity. The wires, made of copper or other good conducting metal, are separately insulated with a plastic. In most cables and extension cords, a third wire for grounding is laid next to the first two and the bundle is covered by another layer of insulation.

Electric current flows through the two inner wires. One is connected to the "hot" or higher voltage side of the transformer (which is outside on the ground, or mounted on a utility pole) and other is connected to the "center tap" of the transformer. The flow of electrons through the circuit powers equipment.

The ground wire protects people and equipment from electrical shocks. If, for example, the hot side wire lost its insulation and made contact with the computer case, you could get shocked by touching the case. By having a ground wire that is attached to the case, the electric current would take the path of least resistance through the wire to the ground rather than through you.

Materials used in wires have free electrons in the outer shells of their atoms. These electrons can move more easily than those in non-conducting materials. An electric force applied to one end of a wire can easily cause an electric current to flow through conductors provided that the other end has a different force.

Monitor Cable

BEHAVIOR

Carries the video signals from the computer to the monitor.

HABITAT

This cable hides in the dusty recesses behind the monitor.

HOW IT WORKS

Typical video graphics array (VGA) monitors use a 15-pin connector that is shielded to prevent electrical interference from other cables and signals. The monitor cable is named the HD-15 cable for high density, 15 pin.

Each of the three colors (red, green, and blue) has its own pins and wire inside the cable. Each color also has a return connection. Two of the other pins are used for vertical and horizontal synchronization, and one pin carries a time signal.

Keyboard Cable

BEHAVIOR

Carries electrical signals between a keyboard and computer.

HABITAT

Connects to the back of a keyboard and plugs into the back of a computer.

HOW IT WORKS

There are several ways to connect keyboards to computers. The oldest way is the plug and cable shown here, the P/S2. There are two configurations of this cable/plug: one with five pins and one with six pins. Adapters can convert between five- and six-pin connections. In addition to the pins, there is a metal tab inside the cylindrical housing that aligns the plug with the socket in the back of the computer.

Newer computers generally connect keyboards to computers either through infrared links (wireless connections using light) or with a USB cable (see page 153).

Serial Cable

BEHAVIOR
Connects two electronic devices.

HABITAT
Inserted into the back of a computer and a PDA docking station, or other device.

HOW IT WORKS
Serial cables transfer information between two devices using RS-232 connectors and nine wires. RS stands for "recommended standard." The original use for this cable and connector was to connect teletypes with modems.

In serial connections, data is sent as a linear series of bits. One bit of the data follows behind the previous bit. This provides a slower rate of data transfer. Faster are parallel connections in which each bit of a string of data is transmitted on separate wires within the cable at the same time. While the serial cable has nine pins, a parallel cable has at least 25.

As computers replaced teletypes, designers kept the same connectors. Today serial cables are used less frequently; new computers usually don't have serial ports. USB connections are used in place of serial cables. Look for this type of cable to become a relic of offices past.

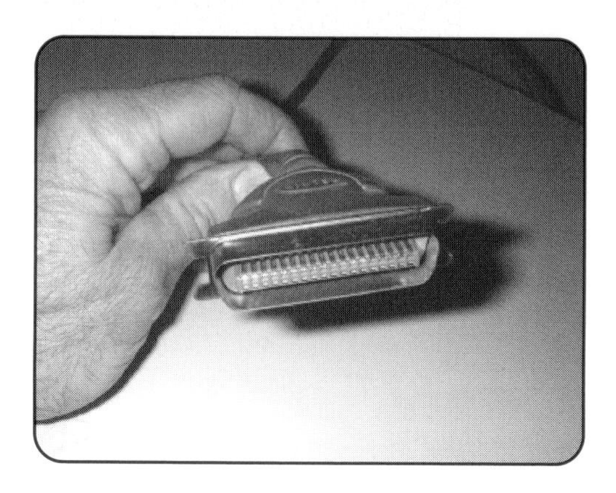

Parallel Cable

BEHAVIOR

Provides higher speed connections between electronic devices.

HABITAT

Usually resides between a computer and a printer.

HOW IT WORKS

Data is transmitted in parallel over several of the wires within the thick bundle of the parallel cable. By transmitting data bytes simultaneously, signals can be transferred much more quickly than through a serial connection. However, parallel cables are dumb—they have no internal control circuits and require the computer to have a driver or driver card so it can communicate with the printer or other device.

New computers may not have parallel ports. USB connections are replacing parallel connections.

USB Cable

BEHAVIOR

Allows electronic devices to exchange data.

HABITAT

USB cables connect computers to keyboards, mice, printers, and games.

HOW IT WORKS

Universal serial bus (USB) cables allow "plug and play" capability. Prior to USB adoption, many devices had to connect to a computer through an expansion card installed in the computer. Thus to add a new device, you would have to purchase an expansion card, open the computer, insert the card into an available slot, put the cover back on the computer, and start it. With USB, you plug in the connector and the computer recognizes what the device is and applies a stored driver to allow immediate use. USB has become the most common connector because of its low cost, its high rate of data transfer, and its plug and play capability.

The USB cable is a twisted pair of conductors—twisted to reduce interference from signals traveling between adjacent conductors. Maximum voltage on the cables is less than four volts.

UNIQUE CHARACTERISTICS

The connectors are asymmetric so they can be inserted only in the correct orientation.

USB stands for universal serial bus. A bus is a device that transfers power or data between different components of a system. Long before computer engineers used the term *bus*, electrical engineers used it for systems that provided power to components.

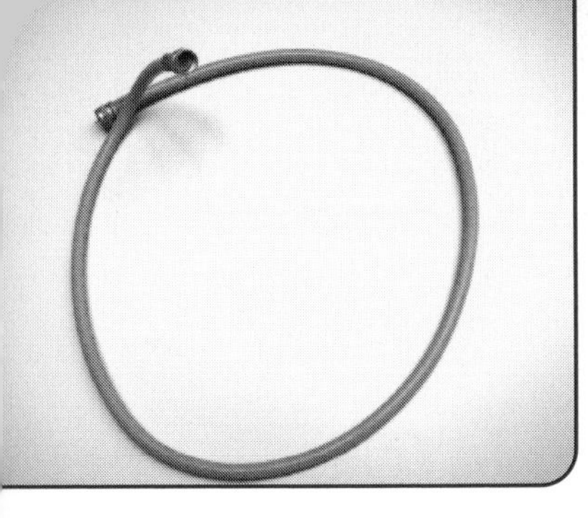

Coaxial Cable

BEHAVIOR
Connects video devices, such as VCRs to televisions.

HABITAT
Found where the cable system cable enters the office and between devices that process video signals.

HOW IT WORKS
Coaxial cables are used to transmit radio frequency signals, usually video signals, with a minimum of interference. The signals are carried on the innermost conductor, and a surrounding shield blocks interference with other nearby cables.

Video signals require higher frequencies and this special cable. Video needs wide bandwidth, which requires the signals to be transmitted at high frequencies. High frequency signals propagate electromagnetic waves outside the wire that can create electric currents in conductors, and these in turn can interfere with the original signal. To eliminate these problems, the signal is confined to the cable by surrounding the signal conductor with the ground wire and insulation.

UNIQUE CHARACTERISTICS
Coaxial cables are easy to spot. Unlike other cables that are rectangular in cross section, coaxial cables are round. Typically they are stiff. They have a round radio frequency (RF) connector or barrel connector at each end; most common is the BNC connector, a bayonet-type connector.

INTERESTING FACTS
Coaxial cables are a cable inside a cable. The inner cable is often a single strand of conductor. Surrounding it is an insulating layer, and surrounding this layer is a braided wire sheath. A final plastic insulator covers the sheath.

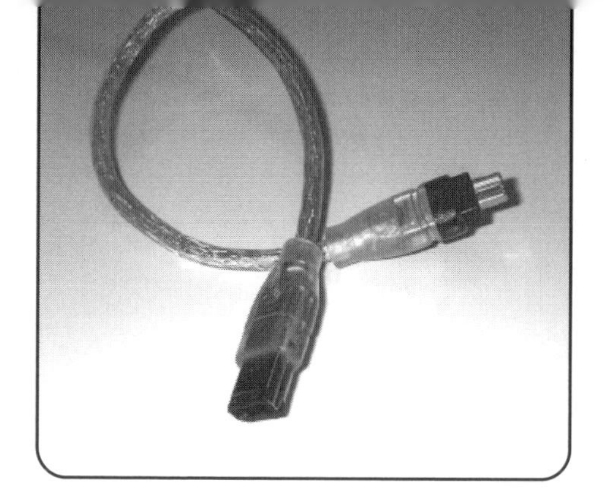

FireWire

BEHAVIOR

Transmits data between video cameras and computers and between other pairs of devices sharing large quantities of data.

HABITAT

Plugged into a computer, waiting to download a camera.

HOW IT WORKS

Although more expensive than USB (because of patent royalties), FireWire provides a faster way to transfer data. FireWire is the Apple brand name for the IEEE 1934 interface. Where the requirement is to transfer large blocks of data quickly, as in video signal processing, FireWire is preferred over USB.

FireWire also allows two devices to exchange data without a computer to moderate the exchange, as is needed with USB protocol. It can also carry electrical power for devices. Typical configurations are six or nine wires.

RJ-45 Cable Connector

BEHAVIOR

Allows the easy connection of Ethernet or CAT 5 cables.

HABITAT

At the end of cables that provide high-speed Internet connections to computers.

HOW IT WORKS

The RJ-45 provides eight pins for for Ethernet or ADSL connections. Ethernet is a system for connecting computers through Local Area Networks (LANs).

CAT 5 cables have four twisted pairs of wires. The cables are twisted to reduce "cross talk" or electrical interference from the signals being sent on other cables.

The RJ-45 looks like the standard telephone jack, RJ-11, but is larger. Despite your best attempt, it won't fit into an RJ-11 socket. Laptops have RJ-45 connectors so you can connect to broadband hubs or routers.

ON WINDOWS

A MAJOR WORKPLACE IRONY is that the most important people have offices with windows, but they never have time to look out of them. How about giving us slackers a view! We'd be more than happy to stare out the windows to make great use of them.

Windows are a building manager's nightmare. They let heat escape in the cold winter and let the sun shine in to heat the office in summer. Plus, they always need cleaning. On the south side of the building where they let in solar heat, workers want the rooms cooler. On the north side of the building, heat radiates and conducts outside to the colder environment in winter. Workers on this side complain about the cold.

Most workers would like to be able to open the windows to let in fresh air and control the room temperature. However, blasts of cold or hot air coming in from outside mess up the central heating system, so building managers usually keep the windows shut. For workers, windows bring sunshine and weather reports, interesting views of people walking by, and eternal hope for vacations or at least the upcoming weekend. They are a luxury, which is why they were taxed in England starting in the seventeenth century.

At one time windows were made from blown glass. Modern plate glass was invented in the early twentieth century and has been improved several times since. Safety glass, made by laminating celluloid between two sheets of glass, was invented in 1910. Today, innovation continues to improve glass, and particularly windows. Windows can change from letting in light to blocking light with the flip of a switch or a command of a computer. Windows may yet please both building managers and occupants.

One-Way Glass

BEHAVIOR

Allows people on one side of the glass to see through into the adjoining room, but prohibits people in the adjoining room from seeing out.

HABITAT

You can find one-way glass in rooms where interviews (employment, coaching skills, and criminal investigation) are conducted. People being interviewed are on one side of the glass, and unseen onlookers are on the other side in an adjacent room. This type of glass is also used in public areas and in stores where managers and security people can watch people without letting them know that they are being watched.

HOW IT WORKS

One-way glass is a sheet of glass coated on one side with a thin layer of reflective metal. The metal layer is only a few molecules thick. This glass is called a half-silvered mirror because it contains only about half the number of metal molecules to block all the light. A fully silvered mirror reflects nearly all the light; a half-silvered mirror reflects about half of the light.

Standing in a well-lighted office, half of the light reflecting off people and objects passes through the one-way glass into the adjacent room, and the other half reflects back. A person on this side of the glass sees his or her own reflection.

The observation room on the other side is kept dark. Although the light there also reflects and passes through the one-way glass, since light levels are so low those images are overwhelmed by the images from the much brighter room. Turn up the lights in the darkened room and you can see through equally well in both directions.

Smart Window

BEHAVIOR

Changes from transparency to translucency with the flip of a switch. If you want to see outside one minute but want privacy the next, flip the switch. If the sun shining through the window is making your guests uncomfortable, flip the switch to block the glare.

HABITAT

Used only in expensive homes and office buildings.

HOW IT WORKS

Smart windows are made of electrochromic, suspended particle devices, or liquid crystal panes that change colors or transparency when an electric charge is applied to them.

Electrochromic windows change color depending on the imposed electric current. With no current applied, they are transparent and let light pass through. As increasing levels of current are applied, the windows change color in a chemical reaction and increasingly block light. The electric current causes lithium ions to move from one electrode to another between two sheets of glass. With electric current applied, these ions absorb light. When the current is turned off, the ions chemically bond and lose both their electric charge and their ability to absorb light. An advantage of this system is that the window can block the sun's glare and heat without blocking the view. The two photos above show the same window with and without the electric current.

Suspended particle devices are sandwiched in a film between two sheets of glass. When a small electric current is applied through a transparent conductor, the particles line up to allow light to pass through the window. With no current, they block all the light.

As the current is decreased, the particles tend to reorient themselves randomly and block some of the light. Liquid crystal windows can be transparent or translucent.

Smart Window Blind

BEHAVIOR

These blinds open and close automatically, depending on ambient temperatures, or are operated manually to let in the sun's rays or block them out.

HABITAT

Mounted in window frames of nicer offices. They are especially useful on windows that are inaccessible.

HOW IT WORKS

Less technically innovative than smart windows, smart blinds operate by remote control. Thermostats, timers, or people can flip a switch to let sunlight in or keep it out. The controller operates a motor that opens and closes the blinds.

Perimeter Entry Detector or Burglar Alarm

BEHAVIOR

It alerts you or a security company that someone has entered the office through a window.

HABITAT

These detectors are affixed to window panes with wire leads running along the window frame or into the wall.

HOW IT WORKS

There are several types of alarm detectors for windows. If the windows open, a magnetic reed switch is often used. A magnet is attached to the window and a magnetic reed switch is attached to the window frame, adjacent to the magnet. When the window is opened (assuming that the alarm has been set), the magnet moves and closes the reed switch to set off the alarm.

Other detectors sense if the glass has been broken. Older detectors were conducting wires embedded in the glass. If the window was broken, so was the conduction between the wires, which would sound the alarm. Sometimes motion sensors are attached to glass, so if the glass is shattered the sensor will move to set off the alarm.

Global Positioning System (GPS) Receiver

BEHAVIOR

It receives signals from several GPS satellites and sends the signals, via a cable, to a computer and transmitter.

HABITAT

This receiver is attached to a window that faces south (toward the equator, where the satellites are) and is out of the way. Once installed, it needs no maintenance, so it can be positioned out of sight and reach.

HOW IT WORKS

Why do you need a GPS receiver in an office? The GPS system works by measuring time differences between various satellites, each of which must have exact time. The office system takes advantage of having exact time broadcast by passing satellites to set all the clocks within a building.

The window-mounted receiver picks up time signals from the satellites and relays them to a radio transmitter that sends signals to all the clocks within the building. Using this system to send out synchronizing signals to the clocks is more cost effective than hard wiring all the clocks or manually setting them. GPS receivers can also trigger bells to indicate lunch breaks or the end of the day.

UNIQUE CHARACTERISTICS

When time shifts between daylight savings time and standard time, the transmitter makes the switch and sends out a signal. If you're standing in front of a clock at 2 A.M. on the equinox in spring or fall, you'll see the clock hands speed ahead or speed backward to adjust to the proper time.

IN THE KITCHEN AND BATHROOMS

OFFICES AREN'T JUST a bunch of desks. Many have kitchen areas and bathrooms as well. All are important to the functioning of the business. Some of the dot-comers have (or at least *had*) a variety of other special places: ping-pong rooms, exercise rooms, and game rooms. Being on the dark side of the dot-com bust, we'll skip these and focus on the more common spaces and the technology they hold.

Coffeemaker

BEHAVIOR

Transforms mere water and roasted beans into a stimulant that drives the nation.

HABITAT

On kitchen countertops and on people's desks in some offices.

HOW IT WORKS

There are several different types of machines. The type shown above (drip) has you pour fresh water into a reservoir and add ground coffee beans into a container that is on top of the coffeepot. Turning on the power sends electricity through a heating element that curves around the inside of the base, below the coffeepot. The heating element is connected to a tube through which the water passes in one direction. The heating element heats the water in the tube, eventually boiling the water. Resulting steam bubbles rise up a tube, pushing hot water ahead into the container with the ground beans. The water drips down onto the ground beans and collects in the pot below. The hot water frees the oil in the beans so the aroma and flavor wash into the pot below. A filter retains the grounds while letting the liquid pass.

In the coffee press, grounds are flooded with boiling water and later filtered out. Percolators repeatedly send water through the grounds and back to the pot. A well in the bottom collects water or coffee, heats it to a boil and sends it up the hollow tube. It spurts out the end, hitting the viewing cap, and drains down through the coffee grounds and into the pot below. Espresso is made by using pressure to force hot water through the grounds.

The story (or myth) of coffee's discovery is that when a goat herder in Ethiopia noticed goats eating coffee beans and saw their unusual vitality, he tried the beans himself. From that simple beginning we now have a Starbucks on every corner.

Microwave Oven

BEHAVIOR

Heats up the cold burrito and the lukewarm cup of coffee to get you through the all-night report or proposal writing.

HABITAT

The kitchen area of an office.

HOW IT WORKS

The oven uses electric power to generate microwave radiation with a vacuum tube called a "magnetron." A stirrer rotates above the cooking food to distribute the microwaves throughout the oven. It rotates about once a second, and you can see it (or its shadow) slowly spinning when you open the oven door. The radiation has a wavelength of about 12 cm (5 inches) and can pass through glass and plastic much like visible light can pass through a window. So the glass cover on a food dish doesn't interfere with the microwaves heating the food, but it helps to retain the heat and cook the food. To keep you safe from the microwaves, the oven door has a latch that prevents it from opening when the microwaves are being generated. The window in the door has a metal screen to keep the microwaves from escaping through the glass. The screen mesh on the inside of the glass has openings much smaller than the wavelength of the radiation and thus blocks the escape of radiation, but allows visible light through. So you can see the cheese on the burrito bubbling.

Water, fats, and sugars in food absorb microwave radiation. The water, fat, and sugar molecules vibrate back and forth as they try to align themselves with the fast-changing electric field set up by the microwaves. This motion generates the heat that cooks the food.

UNIQUE CHARACTERISTICS

Microwave ovens are energy efficient machines. In traditional ovens 50 percent or less of the heat generated heats the food, but microwave ovens operate at 70 to 80 percent efficiency.

INTERESTING FACTS

Engineer Percy Spencer discovered the cooking effect of microwaves by accident. The story is that while experimenting with microwaves, he discovered that a candy bar in his pocket had melted. When he associated the melting with the microwave radiation, he first tried popping popcorn. When that worked, he made the world's first microwave mistake—trying to cook a raw egg. It exploded, and Percy's mind exploded with the concept of microwave cooking.

Automatic Light Switch

BEHAVIOR

Transforms the darkened room into a lighted space upon entering. Conserves energy by keeping the lights off when the room isn't being used.

HABITAT

Frequently installed in bathrooms, the sensor is mounted on the wall near the entry door, usually where you would expect to find a manual switch. In some cases building managers replace manual switches with automatic switches at the same location.

HOW IT WORKS

There are a variety of approaches, but a common one uses passive infrared detectors. These detect infrared radiation given off by a person entering the bathroom and switch on the lights through a relay switch. When the detectors no longer detect a warm body, an internal timer counts down a preset time before switching the lights off.

All objects radiate energy. The wavelength of the radiation depends on the temperature of the object. So the sun (9,900 degrees Fahrenheit at its surface) radiates at a different wavelength than do humans (with skin temperatures about 91 degrees Fahrenheit). The wavelengths of human radiation are in the infrared band.

One problem with automatic lighting is when the sensor does not detect the person inside a bathroom stall. The sudden dousing of lights and inability to turn them back on is, at the very least, annoying for the bathroom occupant.

UNIQUE CHARACTERISTICS

Automatic switches reduce electric power usage for lights by as much as 50 percent, depending on bathroom traffic.

Toilet

BEHAVIOR
The sanitary disposal system for biological by-products.

HABITAT
The anchor plumbing fixture in bathrooms. In other countries toilets are located in "water closets" and bathrooms are for reserved for taking baths.

HOW IT WORKS
Think you know? It's not as simple as you may think. The toilet is a marvel of engineering, flush with the success of countless innovators stretching back to Thomas Crapper in 1886.

If the model in your office has a reservoir, lift the lid to peek at the inner workings. Turning the handle lifts the flapper valve which, once lifted, stays in place until all the water has drained out. As the water level drops the ball float lowers, which turns on a valve connected to the end of the float arm. The valve lets fresh water come in from your cold-water pipes.

There is no valve separating the stuff in the toilet bowl from the sewer line. (Hence those weird, but possible, stories about rats, snakes, and who knows what else crawling up a sewer line and into a toilet.) On flushing, the out-rushing water pulls the bowl contents up and over the sill in the bowl. Once over the sill, it falls or flows down the sewer pipe to the wastewater treatment plant.

Once the reservoir water has emptied, the flapper falls shut and the in-rushing water refills the tank. As the water level rises, the ball float rises and the float arm shuts the valve letting in water. Another toilet type uses water directly from the water main. Pulling the flush lever or pressing the button opens a valve that lets in the water. Air pressure closes the valve after a preset time. This is called a "flushometer" system.

Another system uses air pressure to help remove waste from the bowl. Air is pressurized by the water pressure in the main. On flushing, the air forces the water into the bowl at a much faster speed than traditional toilets. This pressure-assist toilet saves water and expense.

Flushless Urinal

BEHAVIOR

Saves water by not cleaning the bowl after every use.

HABITAT

In newer office bathrooms, especially in regions where water conservation is especially needed.

HOW IT WORKS

Urine collects in the pipe below the bowl and washes out the bend in the drain pipe as more volume collects.

In most urinals, water washes down the bowl and pushes urine out the drain to eliminate odors from urine in the pipes. Flushless or waterless urinals eliminate the water pipes, covering the collected urine with a thin film of liquid. The liquid is less dense than water or urine and floats on the surface, blocking odors. In one design, the urine floats up through the lighter liquid and flows into the drain, leaving the liquid in place.

The flushless system conserves water (one company claims each unit can save 45,000 gallons per year) but requires replacing the liquid after 1,500 uses.

INTERESTING FACTS

Urinals aren't often the topic of civic controversy, but in some cities they now are. Conservationists encourage the installation of waterless urinals, and some health officials recommend them because they eliminate splatter and have no dirty buttons to push. However, many plumbers oppose them. Stay tuned to find out how this all flushes out.

Automatic Flush Toilet and Urinal

BEHAVIOR

Flushes the unit when the user leaves.

HABITAT

In office bathrooms, mounted on the wall behind the toilet or urinal or attached to exposed piping.

HOW IT WORKS

Automatic flushing systems use either an active or passive sensor, and some have a manual mode as well. An active sensor with an infrared radiation beam transmits an infrared burst that reflects off a user when he or she approaches. When the beam is no longer reflected back to the sensor, indicating that the user has left, the sensor triggers a solenoid that controls the water valve. In passive systems, the sensor does not send out a beam to be reflected, but instead it picks up the body's infrared energy.

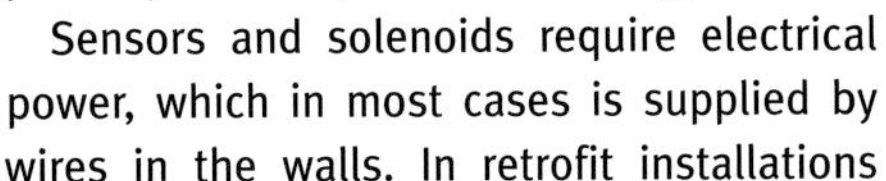

Sensors and solenoids require electrical power, which in most cases is supplied by wires in the walls. In retrofit installations where the sensor is not mounted into the wall, replaceable batteries are required.

Some automatic flushers don't use sensors and instead operate periodically, controlled by a timer.

INTERESTING FACTS

Why go to the expense of installing automatic flushers? Surprisingly, many users don't flush in public restrooms, leaving a less sanitary and less pleasant environment for everyone else. And, automatic flushers provide maintenance savings—there are no handles and valves to replace.

Ground Fault Interrupt (GFI) Switch

BEHAVIOR

It automatically interrupts electrical power when it senses a surge in current.

HABITAT

GFI switches are most often located adjacent to sinks in bathrooms or kitchen areas.

HOW IT WORKS

A GFI switch, sometimes called a Ground Fault Circuit Interrupter (GFCI), shuts off the electric flow when you might be the electrical conductor between an appliance and ground. How could that occur? If you're making coffee and the electric pot is not grounded, electricity could flow from the pot through you to the cold-water pipe when you turn on the water. The cold-water pipe acts as a ground, and if you are holding a hot wire or faulty appliance with one hand and a ground with the other, you will be shocked. The GFI switch senses a sudden surge in current and shuts off.

Two coils inside the switch sense the flow of current through the hot terminal and the neutral terminal. If there is an imbalance between their current flows, it shuts the circuit within a few milliseconds.

INTERESTING FACTS

Some industry experts estimate that as many as 200 electrocutions a year could be prevented if everyone used GFI switches when operating electrical appliances near cold-water pipes.

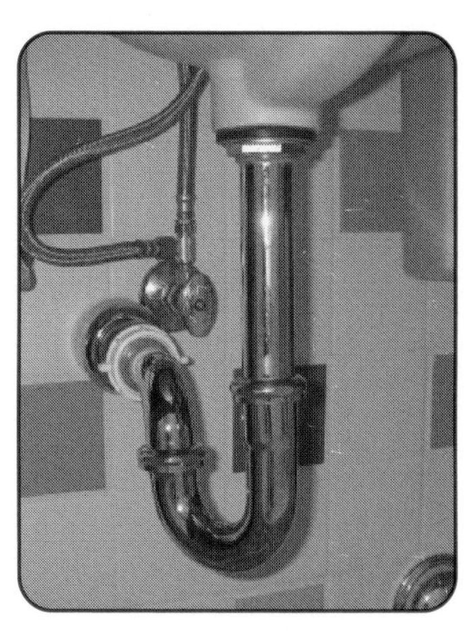

P-Trap

BEHAVIOR
Prevents sewer gases from entering the office through sinks and drains.

HABITAT
Beneath every drain is a P-trap.

HOW IT WORKS
The P-trap, which looks more like the letter U, traps wastewater in the bottom of the U. As wastewater flows out of a sink, it displaces the water that had been occupying the trap. The trap keeps water in the U to prevent gases, generated by bacteria in wastes flushed down the drain, from coming up through the sink and into the room.

UNIQUE CHARACTERISTICS
At the bottom of most P-traps is a large nut or cleanout plug. When the P-trap becomes clogged, you call the building maintenance guy to clean it out (what a fun job this is!) by opening the cleanout plug with a large wrench.

INTERESTING FACTS
Drain parts tend to come in three colors: white (made of PVC plastic), black (made of ABS plastic), and silver (metal).

Hand Dryer

BEHAVIOR

Dries hands without a towel by blowing dry air onto hands.

HABITAT

In bathrooms adjacent to sinks where paper towels or continuous rolls of linen towels are not provided

HOW IT WORKS

The traditional model requires the user to punch a button to start the flow of warm air. An electric heating element inside warms the air blown through by a fan blade mounted on an electric motor. The heating element is made of highly resistive wire through which electric current is passed after the switch has been pushed. A timer turns off the heater and fan.

A newer model commonly used in Japan (and now available in the United States) called Jet Towel promises to be more effective. You put your hands inside the unit and a sensor detects them and turns on the blower. Rather than trying to evaporate the moisture from your hands with warm air, this machine blows the water away with a hurricane-strength blast of air. In five seconds your hands are dry.

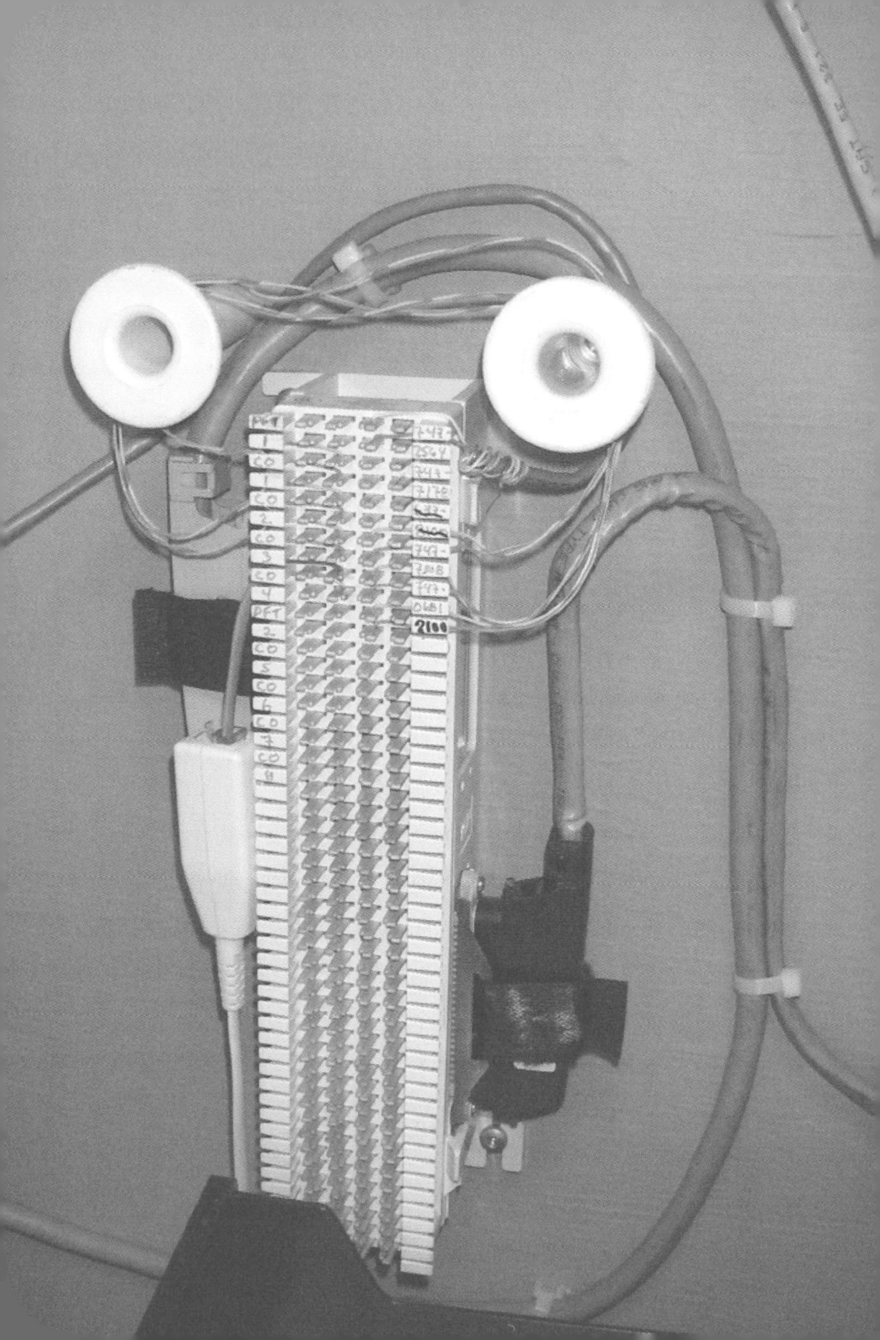

11 BEHIND THE SCENES

DO YOU LIKE EXPLORING dark recesses? It's time for a tour of those places in the office you rarely see but are just a few steps away from your desk.

Controls, switches, and connections for all the stuff you use while you work have to be located somewhere, usually out of sight. Controls for the flow of the air, outside and inside connections that let you make telephone calls, and machines that provide access to the Internet all reside behind the scenes.

Grab this book and a flashlight and go exploring!

Registered Jack, Style 21X (RJ-21X)

BEHAVIOR

This connecting block or jack connects the outside telephone lines, coming from the telephone service provider, to the lines or switches in the office. It is the interface connection between the phone company's responsibility and the office's responsibility.

HABITAT

This connection is mounted on a wall, usually in a closet out of sight. However, the one shown here is mounted on the office wall visible to everyone.

HOW IT WORKS

The installer pushes the insulated wires into the block where the contacts press through the wires' insulation, establishing an electrical connection. The installer then cuts off the ends of the wires to complete the connection.

Wires leading to phones on desks or to telephone switches are inserted into corresponding blocks to make connections. The jack provides an easy way to connect the inside and outside lines.

Typically this type of connecting block is used in small businesses or small offices. You won't find it in homes or in large companies.

UNIQUE CHARACTERISTICS

In the photo you can see the telephone numbers of the incoming lines printed on the right side. It appears that this office has five outside lines.

Patch Panel

BEHAVIOR

A patch panel allows for easy connection to computers in the office.

HABITAT

Patch panels are usually mounted on walls in out-of-the-way places, often in closets and away from prying fingers.

HOW IT WORKS

These panels connect computers to the Internet. A RJ-45 plug (eight-connector plug) plugs into the panel. The plug is the connector for an Ethernet cable. At the other end of the cable, another RJ-45 plug inserts into a computer (Network Interface Card). The DSL line connects to the two 48-port station panels (the upper two panels) through the hub (the lowest black panel). The hub distributes the DSL connection to the various computers in the office, so they all can share one DSL line.

The DSL line is a telephone line that comes into the office through the RJ-21X connecting block and goes to a modem. An Ethernet cable takes the signal to the hub for distribution to the patch panels where all the computers are connected.

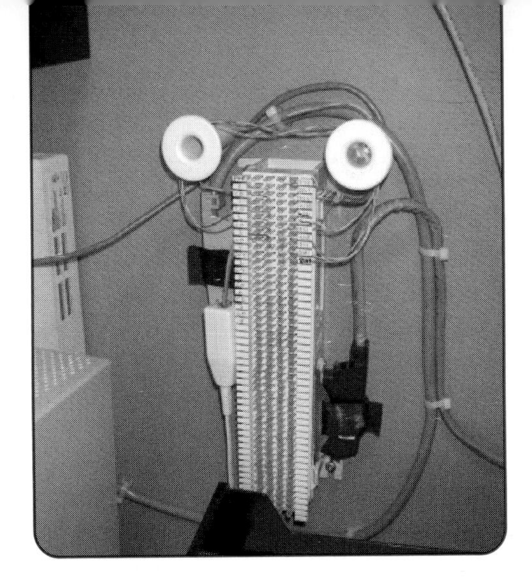

Split 50 or Rolm Block

BEHAVIOR

This block provides quick connection points for telephone lines, modems, telephone sets, paging equipment, and fax lines.

HABITAT

Mounted on a wall near the other telephone equipment, often in a closet.

HOW IT WORKS

The block provides connection points to all the devices that connect to the telephone switch or private branch exchange (PBX).

The Split 50 has 50 pins on each side of the block that accommodate 25 pairs of wires. Wires leading to telephones plug into the right side, and wires leading to the PBX plug into the left side. The left and right are not internally connected so the installer has to wire them together.

UNIQUE CHARACTERISTICS

On the left side in the photograph is a DSL filter. This is a low-pass filter that allows conversational frequency range signals to pass, but not the higher frequency signals (used for Internet service) that could interfere with phone conversations. Human conversation occurs in the hundreds to thousands of cycles per second frequency range (0 to 4,000 hertz, or cycles per second). DSL lines use much higher frequencies on the same line for transmitting data to computers. The filter keeps the higher frequency DSL signals from phone lines.

INTERESTING FACTS

One of the great innovations of our age has been the increase in data transmission capacity. When we think of the Internet age we may think of software (browsers, Web sites, and so on) and fast computers. All that speed would go for naught without transmission capacity. In the old pre-Internet days one telephone conversation needed one pair of twisted wires dedicated to it continuously. Now, that single pair of wires can carry many phone conversations, Web surfing, and faxing at the same time. The enhanced transmission capacity is an innovation of using higher frequency bands and sending data digitally in packets rather than as analog signals. See the VOIP Headset entry (p. 141) for a description of packets.

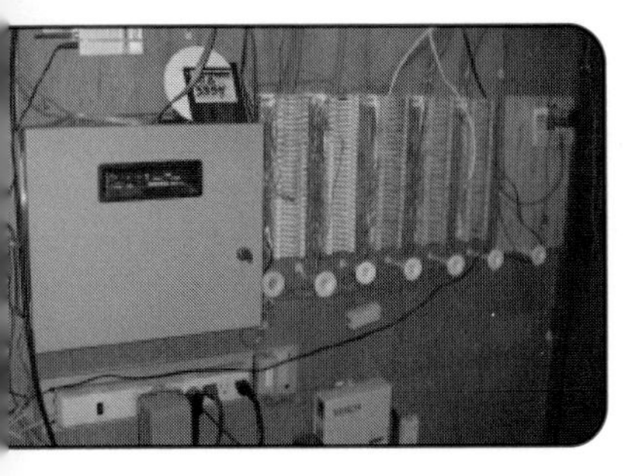

PBX

BEHAVIOR
It sends your phone calls to the telephone number you've dialed.

HABITAT
Mounted in an equipment rack in a closet or located off-site at the telephone company's local exchange.

HOW IT WORKS
Private branch exchanges (PBXs) provide myriad services, but most important among them is that they connect two or more telephones. For outgoing calls, the PBX selects an available telephone line and connects the user to it automatically.

A PBX makes the connection so you can call someone else in your office, and makes the connection to an outside line so you can call anyone outside your office. It also can provide background music for calls on hold, call forwarding, voice mail, speed dialing, conference calling, and accounting of calls.

Picking up your phone sends a signal to the PBX, and it responds by turning on the dial tone. As you enter the phone number the PBS either routes the call to the internal phone you're calling or to a "trunk" line that connects you to the switches in the telephone company's central office.

A PBX saves a company from having to connect every phone in the office to an outside line and paying all the line fees when most phones are used only a fraction of the day. It also allows people inside the office to call each other.

UNIQUE CHARACTERISTICS
If your office phone requires you to dial "9" to get an outside number, your call is being routed through a PBX. On the other hand, if you select an outside line to use for making a call by pushing a button on the phone, you are not using a PBX. This is called a "key system."

Cable Amplifier and Splitter

BEHAVIOR

Strengthens the cable signal so it can be distributed to many different outlets throughout the office.

HABITAT

Mounted on a wall in the equipment closet.

HOW IT WORKS

If you have only two or three offices that need cable access, you don't have one of these devices. However, if you have cable in several offices, you might.

The incoming cable signal is only strong enough to provide access to a small number of outlets. To supply more, the signal needs to be amplified. Once strengthened, it can be split among several outlets.

In the photo, the amplifier is the black box. Above the box are some of the splitters that divide the signal among the cable outlets.

Network Server

BEHAVIOR

Sends out company Web pages to everyone who types the company URL into their browser.

HABITAT

In equipment rooms, in large racks with other servers. In smaller companies, the server may be a PC sitting on the floor.

HOW IT WORKS

A server is a computer that provides specific information or service to other computers. A Web server provides Web pages to requesting computers.

Typing a URL into your browser starts a chain of events that result in the Web page showing up on your screen. You enter three pieces of information into your browser to pull up a Web page: the protocol (for Web pages, it's *http*; for file transfer protocol, it's *ftp*), the server name (the part of the address between the "//" and the "/" if there is one in the address), and the file name (comes after the "/" following the server name). The browser sends a request for the desired file to the server using the specified protocol.

However, it's not that simple. Another server in the system interprets the domain name or server name into its actual Internet Protocol (IP) address. An IP address is a 32-bit number that is unique to each server on the Internet. This "name server" keeps a table of domain names and IP addresses.

Now the request for a Web page is heading to the proper server. The Web server retrieves the requested file from its hard drive and sends it to your browser (in Internet format, packets). Other types of servers include FTP and e-mail servers.

Global Positioning System (GPS) Clock Transmitter

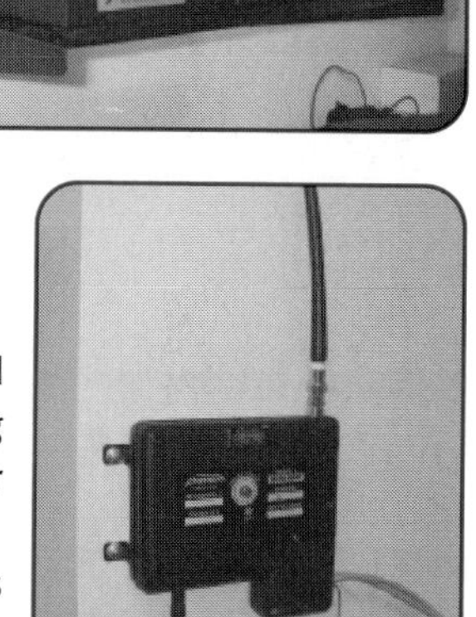

BEHAVIOR

It controls the clocks and time-controlled bells or buzzers in a building.

HABITAT

This radio transmitter is housed in an equipment closet, probably mounted on a wall.

HOW IT WORKS

You might wonder why someone would want a navigational device inside a building This type of GPS system isn't used for navigation; it supplies the correct time.

The transmitter gets its timing signals from a receiver mounted on an outside window that receives GPS signals from passing satellites. The transmitter decodes the signal and sends radio signals to all the clocks in the building, directing their motors to speed up or slow down. In large offices that have many clocks, such systems save confusion and money by synchronizing all the clocks.

If an office has a time-controlled bell or buzzer system, the transmitter sends radio signals to a separate device that rings the bells at the programmed time.

UNIQUE CHARACTERISTICS

Follow the wires in the equipment closet that are attached to the transmitter. One will lead probably upward, toward the ceiling, to connect to the window-mounted receiver. If the office has bells, a second wire will lead to the device that controls them. A third wire will supply electrical power to the transmitter.

Pneumatic Controls for HVAC

BEHAVIOR

Opens and closes valves that allow air to enter the HVAC system.

HABITAT

These controls are in equipment closets of large office buildings. Since they have to be protected from tampering, they are housed behind locked doors.

HOW IT WORKS

HVAC (heating, ventilation, and air conditioning) systems include thermostats to sense temperature, ducting to carry warm or cool air, pumps to move the air, heaters and air conditioning units, and valves or switches.

Just as engineers can use electronic controls to open valves to let air pass, they can also use pneumatic controls. This older system relies on pressure differences instead of voltage or current differences to transmit control information.

ON THE MOVE

STAND IN LINE at an airport and you won't believe reports that business travel is on the decline. The argument goes that Web conferencing and teleconferencing are replacing business travel, but crowded airports and planes suggest differently.

Business travel is a way of life for many people. Road warriors hit the pavement and fly the skies to sell, consult, and manage. And while they are enjoying those airline meals or non-meals and lonely evenings in hotel rooms, the work at home is piling up. That's why productive road warriors take their offices with them.

A quick stop at the office supply store or a few minutes surfing the Web turns up enough hardware to keep the traveling executive connected and working 24/7. The miniaturization of electronics has produced computers and gadgets that allow us to develop heartburn anywhere in the world as bad news and new requirements hit us at the new speed of business. Ah, the glamour of travel!

Cell Phone

BEHAVIOR
Allows users to be interrupted at all times and in all places. And allows them to interrupt other people at all times from all places.

HABITAT
Usually resides in pockets, pocketbooks, briefcases, or belt holsters. Sometimes it is kept on office desks or in cars.

HOW IT WORKS
The marvel of cell phones is that computers can track users and send incoming calls to them anywhere, and reroute calls as users move throughout an area of coverage. To originate a call from a cell phone, a user dials a number and sends it to a nearby cell tower antenna. There it travels to a switching station that routes the call. The call can be transmitted long distances by microwaves or through optical fiber or copper wires.

As users move, computers in the system switch the path of the call from one cell tower to another for better reception. Users are unaware of the switching as it occurs so quickly their conversation is not interrupted.

Cell phones identify themselves to the local switching station. That station searches for incoming calls and routes them to the nearest cell tower. After your phone rings, it sends a signal to the caller (the sound of the phone ringing) to let him know that the connection is being made.

Cell phones use two radio signals. One carries your voice transmission to the tower, and the other carries the other person's voice to you.

Older cell phones use analog signals, but newer phones are digital. Voice and images are converted into digital signals and are transmitted. Digital phones have higher sound quality due to less electrical interference.

INTERESTING FACTS
Motorola made the first commercial handheld mobile phone in 1983.

Personal Digital Assistant (PDA)

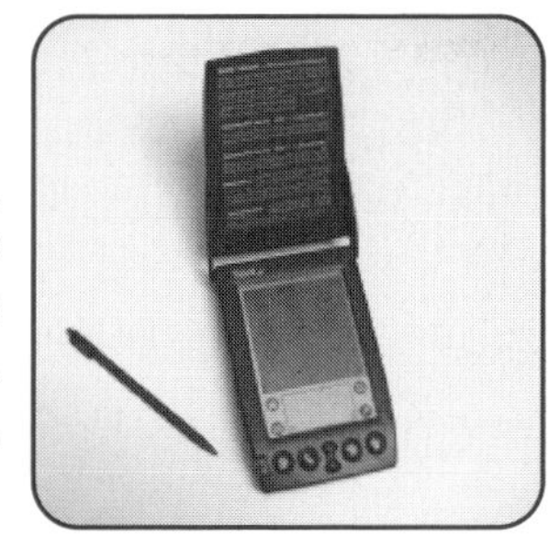

BEHAVIOR

It remembers all those numbers, names, and dates that you used to write in a notebook or calendar. Increasingly PDAs also act as cell phones, music players, e-mail and Internet access providers, and providers of electronic games.

HABITAT

Carried in pockets, purses, briefcases, and backpacks.

HOW IT WORKS

PDAs share information with PCs, so appointments recorded on one machine can be uploaded to the other. Older PDAs require a cable hookup to a computer; newer models use Bluetooth or other wireless technology to share information with PCs.

You enter data and select options either by pressing a stylus to a touch screen or by moving a cursor using a thumb wheel. Some PDAs come with a tiny, but complete, keyboard.

PDAs are sold with a host of software that gives users spreadsheets, word processors, and other capabilities. Additional software can be downloaded to the PDAs from the Internet.

Recently, cell phones have been integrated into PDAs, so a road warrior can reduce his or her load and carry only one device to keep in touch and on time. Increasingly, PDAs allow e-mail correspondence through access to Wi-Fi (wireless local area networks).

A PDA is a small computer built around a microprocessor. It has no hard drive, but instead stores your data in a volatile random access memory (RAM). This memory requires a small, continuous supply of electrical power that is provided by batteries. Some PDAs use flash drive memory instead of or in addition to the RAM. Instead of a separate monitor like a PC, PDAs have built-in LCD screens, many of which operate also as touch screens.

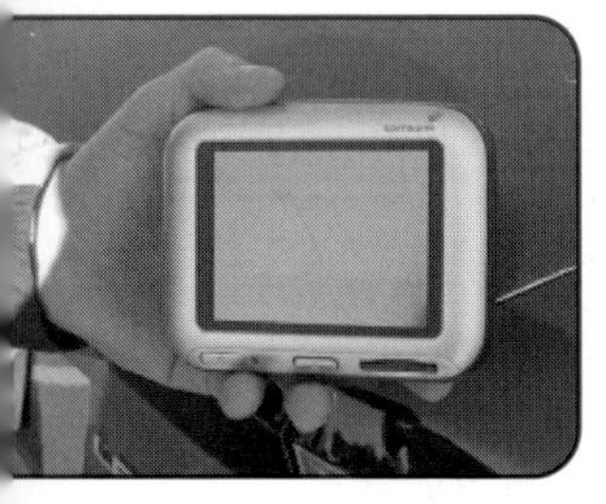

Global Positioning System (GPS)

BEHAVIOR

Guides you to your next appointment. Or, shows you the way to go home.

HABITAT

GPS units can be mounted on the dashboard of a car or carried in a briefcase.

HOW IT WORKS

It's magic! Imagine 24 satellites swirling high above your head, sending simple messages back to earth that a relatively inexpensive device the size of a PDA can interpret to give you your position anywhere on earth. What would Columbus have given for this?

At any time, only a few of the 24 functioning satellites are "visible." A GPS will search for the signals of several satellites, trying to find at least four signals that it can use. Each satellite sends out a unique signal so the receivers can distinguish from which satellite it is receiving a signal. The GPS calculates the position of the satellite based on information in its memory and a time signal it receives from the satellites. Knowing the satellites' positions, it calculates how long it took each signal to travel from the satellite. From the estimate of signal travel time, the GPS can calculate a distance from each satellite. Knowing the distance from several satellites, it can correct for errors and calculate your position. And it does all that in a fraction of a second.

Knowing your position isn't much help for most people. (For example, knowing that as I write this, I'm at 47 degrees 37 minutes 54.9 seconds N, and 122 degrees 6 minutes 7.7 seconds W, probably won't help you find me.) So, navigational GPS units include maps and way-finding software that translate the GPS-provided position into a location on a street map.

INTERESTING FACTS

The twenty-four GPS satellites revolve around the earth twice a day at 12,600 miles above ground.

Power Adapter

BEHAVIOR

Allows users to recharge cell phones, digital cameras, and other electrical appliances when traveling in countries that use different electric plugs and outlets.

HABITAT

Resides in the luggage of the international road warrior and jet-setter.

HOW IT WORKS

These inexpensive adapters merely make connections between the plugs of two or more different electric systems. They do not change the voltage or frequency of the current.

Thus, in the United Kingdom, inserting a power adapter into a wall outlet provides you with a source of power to plug your laptop into. But that power is at 50 cycles per second (instead of 60 cycles per second in the United States) and the voltage is 230 volts (instead of 110 volts).

In the United States electrical plugs have either two flat metal prongs (Type A) or two flat metal prongs and a cylindrical ground prong (Type B). In the UK, electrical plugs use a configuration of three flat metal bars, two parallel and one perpendicular to the others. Much of Europe uses a two round prong system (Type C), and this same configuration with a ground connection in the other rim of the plug (Type F).

Some countries use up to seven different types of plugs and sockets, some of which are Type A or B but deliver a different voltage. So before plugging in when you're overseas, find out what the voltage and frequency are.

UNIQUE CHARACTERISTICS

Look at the appliance and its power supply before plugging it in. It will indicate what it can safely accept in terms of frequency and voltage. Many U.S.-made appliances can be recharged with different power systems.

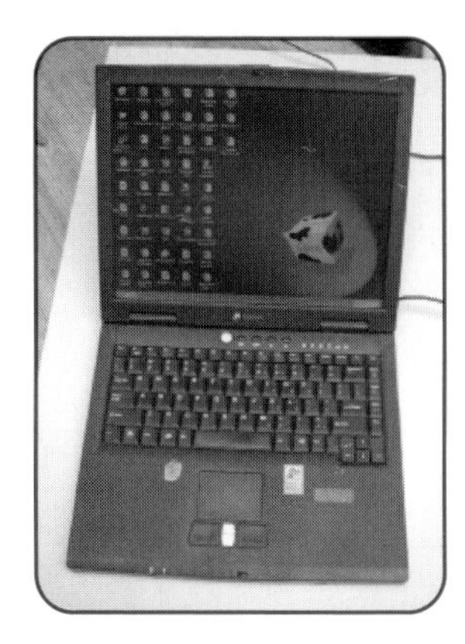

Laptop

BEHAVIOR

Provides computing capability, including access to the Internet, in a lightweight and portable unit.

HABITAT

Laptops sit on desks in place of desktop computers but also reside in carrying cases lugged about by road warriors. Seen often at Starbucks and other wireless hotspots.

HOW IT WORKS

Take off the metal jacket of a desktop computer to reveal how much empty space there is. That space is missing in a laptop. And, laptop components are made smaller and lighter than they are in a desktop computer. The result is that there is a lot of number-crunching power in a very small space.

Laptops have a built-in keyboard and monitor. Laptop monitors are usually liquid crystal displays (LCD) rather than the cathode ray tube monitor of a desktop computer. LCDs are both smaller and lighter, and require less electrical power.

Laptops have an internal battery that can power them for an hour or two, and sometimes up to 15 hours. The battery (usually a lithium ion battery) is recharged with an adapter/transformer that plugs into a wall outlet. The computer can be operated while plugged in.

Instead of a mouse to direct the cursor, laptops have a touch pad or pointing stick. However, users can connect an external mouse through a USB port.

INTERESTING FACTS

The first laptop, the Osborne 1, weighed in at 24 pounds and sported a five-inch display screen. It was introduced in April 1981 and cost nearly $1,800, about the same as a model today that weighs a fraction as much and has many times the computing power and speed.

Wi-Fi Detector

BEHAVIOR

Scouts out any wireless hotspots. Finds hotspots without having to start up your computer. Some models also find Bluetooth and microwave signals.

HABITAT

Connected to key chains on laptop cases or carried in pockets until needed.

HOW IT WORKS

The Wi-Fi detector listens for Wi-Fi signals on the assigned frequency (2.4 GHz). Weak signals are indicated by one set of LEDs, and stronger signals are indicated with additional sets of LEDs. With this gizmo, you can pick your coffeeshop or restaurant based on the strength of its signal.

Memory Stick

BEHAVIOR

It is a compact form of memory that allows users to transfer information from one computer to another. It is also the geek equivalent of a necklace.

HABITAT

Worn on lanyards around the necks of the ubertechies, this memory device is far more visible than traditional devices like floppy disks and hard drives. Others carry their memory stick in a briefcase, pocketbook, or backpack. During use, they are plugged into USB drives on computers.

HOW IT WORKS

Memory sticks have flash memory, which is a solid-state, nonvolatile system for storing data. Solid-state means it has no moving parts. Nonvolatile means that, unlike computer RAM, the data remains in place even when power is shut off. Once the data is electrically written, it will remain until erased or written over. Digital cameras, mobile phones, and digital audio players also use flash memory.

Memory sticks have several integrated circuits inside the outer plastic shell. One or two store the data. Another controls the flow of data to the computer. The sticks also have an LED to indicate when the stick is sending or receiving data and an onboard oscillator. They get electrical power from the computer they connect to.

Digital information is stored in microscopic transistors in flash drives. Electric charges are applied to individual cells to store each bit of data. The charges are confined to the cells until erased or written over.

Originally, memory sticks were limited to 128 MB of data. Newer versions will be able to hold 64 GB.

UNIQUE CHARACTERISTICS

Because flash memory can be corrupted by static electricity, memory sticks have plastic caps to protect them. Be sure to keep the caps on when the stick is not in use.

Power Inverter

BEHAVIOR

Changes direct current into alternating current. It allows you to use direct current sources, like your car's electric system, to power computers and other devices that require alternating current.

HABITAT

In cars and briefcases, often near laptop computers.

HOW IT WORKS

Many electric devices require 60 cycles per second alternating current that is delivered through wall outlets. Although this is the electric currency of the realm, it isn't available everywhere. You typically can't find this in cars, on trains, or aboard planes. However, direct current is often available in vehicles, and the power inverter changes the direct current into alternating current.

The inverter has a device that switches the current to opposite sides of a transformer, thus providing a change of polarity needed for alternating current. However, the voltage is usually 12 volts while the appliance needs 110 volts. The transformer changes the voltage to the conventional voltage. Electronic circuits smooth out, or filter, the electric signal.

INTERESTING FACTS

Some uninterruptible power supplies (UPS) use inverters, too. The UPS has a constantly charging battery that supplies direct current power if the alternating current power from a wall outlet is lost. It then uses the stored energy to power the device by first changing the direct current into alternating current.

Scooter

BEHAVIOR

Transports people and small loads over moderate distances quickly.

HABITAT

Used in manufacturing plants, large offices, and airports.

HOW IT WORKS

Not quite as complex as the Segway (see the next entry), the scooter requires humans to provide both the balance and propulsion. A person stands with one foot on the scooter and pushes the ground with the other.

INTERESTING FACTS

The Oslo Airport seems to be a center for scooters. A variety of scooters carry newspapers and magazines to airport stores and move airport officials from one end of the cavernous building to the other. The scooters seem to work well and cost a tiny fraction of the price of a Segway.

Segway

BEHAVIOR

Scoots you around the office or to your next appointment in futuristic style.

HABITAT

You can find Segways in large offices where workers need to move quickly. They are used in warehouses, manufacturing plants, and for home and office delivery.

HOW IT WORKS

Two powerful servomotors drive the Segway's two wheels. The servos receive their input from gyroscopic and tilt sensors. When the rider leans forward, the sensors detect the tilt and direct the wheels to move forward to restore the balance. As long as the rider leans forward, the Segway moves forward. Turning is achieved by a manual twist control that slows one servomotor relative to the other. Newer models may incorporate "lean to turn" capabilities.

Two onboard computers—actually two circuit boards with clusters of microprocessors—collect data from the sensors 100 times each second and send out corrections to the two servomotors. The gyroscopes are not like the ones you have played with; you don't need to pull a string to start the metal disk spinning. These gyroscopes are tiny solid-state devices that detect motion and relay it to the microprocessors.

A governor limits how fast the Segway can travel, with a top speed of 10 to 12 miles per hour, depending on the model. Power is supplied by lithium-ion batteries that can be recharged at standard electrical outlets. With a full charge the maximum range is about 24 miles.

INTERESTING FACTS

Dean Kamen invented the Segway in 2001. Although sales have not yet reached predictions, Segways are becoming more common. They are used for delivering mail, patrolling streets and buildings, and generally getting around.

cincinnati

TIME EQUIPMENT CO.
cincinnati

INDEX

ABOUT THE AUTHOR

Ed Sobey is science educator, author, and former museum director. He has directed five museums including the national museum of inventing: the National Inventors Hall of Fame. While directing A.C. Gilbert's Discovery Village, Ed founded the National Toy Hall of Fame, now at the Strong Museum in New York.

Ed holds an adjunct faculty position at California State University, Fresno, where he developed Kids Invent Toys, a technology-learning program used by universities and museums in North and South America, Australia, and Asia. He is a Fellow of the Explorers Club, having participated in expeditions, research, and exhibition work throughout the world. Ed holds a Ph.D.

A Field Guide to Roadside Technology

BY ED SOBEY

A Selection of the Scientific American Book Club

"Fun, informative, and easy to use." —SCHOOL LIBRARY JOURNAL

"Proves there really is a travel book out there for everyone."
—DENVER POST

If you've surveyed the modern landscape, you've no doubt wondered what all those towers, utility poles, antennas, and other strange, unnatural devices actually do. *A Field Guide to Roadside Technology* is written just for you. More than 150 devices are grouped according to their "habitats"—along highways and roads, near airports, on utility towers, and more—and each includes a clear photo to make recognition easy. Once the "species" is identified, the entry will tell you its "behavior"—what it does—and how it works, in detail. You'll also learn the history and little-known facts behind the devices you might otherwise take for granted.

$14.95 (CAN $18.95)
ISBN-13: 978-1-55652-609-1
ISBN-10: 1-55652-609-1

A Field Guide to Household Technology
BY ED SOBEY

Illustrating how a fire alarm detects smoke and what the "plasma" is in a plasma-screen television, this fascinating handbook explains how everyday household devices function and operate. Over 180 different household technologies are covered, including gadgets unique to apartment buildings and houseboats. Devices are grouped according to their "habitats"—the living room, family room, den, bedroom, kitchen, bathroom, and basement—and features detailed descriptions of what the devices do and how they work, with photographs for easy identification. With helpful sidebars describing related technical issues, such as why a cheap dimmer switch can interfere with radio reception, this handbook for curious readers also provides the history behind many older household technologies like toasters and faucets and the science behind newer technologies like motion detectors, TiVo, and satellite radio.

$14.95 (CAN $18.95)
ISBN-13: 978-1-55652-670-1
ISBN-10: 1-55652-670-9

Backyard Ballistics

Build Potato Cannons, Paper Match Rockets, Cincinnati Fire Kites, Tennis Ball Mortars, and More Dynamite Devices

BY WILLIAM GURSTELLE

A Selection of Quality Paperback Book Club

"Your inner boy will get a bang out of these 13 devices to build and shoot in your own back yard, some of them noisy enough to legally perk up a 4th of July."—THE DALLAS MORNING NEWS

"Would-be rocketeers, take note: Engineer William Gurstelle has written a book for you."—CHICAGO TRIBUNE

"William Gurstelle . . . is the Felix Grucci of potato projectiles!"
—TIME OUT NEW YORK

This step-by-step guide uses inexpensive household or hardware store materials to construct awesome ballistic devices. Features clear instructions, diagrams, and photographs that show how to build projects ranging from the simple—a match-powered rocket—to the more complex—a scale-model, table-top catapult—to the offbeat—a tennis ball cannon. With a strong emphasis on safety, the book also gives tips on troubleshooting, explains the physics behind the projects, and profiles scientists and extraordinary experimenters.

$16.95 (CAN $25.95)
ISBN-13: 978-1-55652-375-5
ISBN-10: 1-55652-375-0

Engineering the City

How Infrastructure Works
Projects and Principles for Beginners
BY MATTHYS LEVY AND RICHARD PANCHYK

"Future engineers, math enthusiasts, and students seeking ideas for science projects will all be fascinated by this book."—BOOKLIST

How does a city obtain water, gas, and electricity? Where do these services come from? How are they transported? The answer is infrastructure, or the inner, and sometimes invisible, workings of the city. *Engineering the City* tells the fascinating story of infrastructure as it developed through history along with the growth of cities. Experiments, games, and construction diagrams show how these structures are built, how they work, and how they affect the environment of the city and the land outside it.

$14.95 (CAN $22.95)
ISBN-13: 978-1-55652-419-6
ISBN-10: 1-55652-419-6

www.chicagoreviewpress.com

Distributed by
Independent Publishers Group
www.ipgbook.com

Available at your local bookstore
or by calling (800) 888-4741